T0205686

# WIND RESOURCES AND FUTURE ENERGY SECURITY

## Environmental, Social, and Economic Issues

# WIND RESOURCES AND FUTURE ENERGY SECURITY

## Environmental, Social, and Economic Issues

*Edited by*
**Muyiwa S. Adaramola, PhD**

| Apple Academic Press Inc. | Apple Academic Press Inc. |
| 3333 Mistwell Crescent | 9 Spinnaker Way |
| Oakville, ON L6L 0A2 | Waretown, NJ 08758 |
| Canada | USA |

©2015 by Apple Academic Press, Inc.

First issued in paperback 2021

*Exclusive worldwide distribution by CRC Press, a member of Taylor & Francis Group*

No claim to original U.S. Government works

ISBN 13: 978-1-77463-558-2 (pbk)
ISBN 13: 978-1-77188-144-9 (hbk)

### Library and Archives Canada Cataloguing in Publication

Wind resources and future energy security: environmental, social, and economic issues / edited by Muyiwa S. Adaramola, PhD.

Includes bibliographical references and index.
ISBN 978-1-77188-144-9 (bound)
1. Wind power--Environmental aspects. 2. Wind power--Social aspects. 3. Wind power--Economic aspects. 4. Energy security. I. Adaramola, Muyiwa, author, editor

| TJ820.W45 2015 | 621.31'2136 | C2015-901403-4 |

### Library of Congress Cataloging-in-Publication Data

Wind resources and future energy security : environmental, social, and economic issues / Muyiwa S. Adaramola, PhD, editor

pages cm
Includes bibliographical references and index.
ISBN 978-1-77188-144-9 (alk. paper)
1. Wind power. 2. Renewable energy sources. I. Adaramola, Muyiwa, editor.

| TJ820.W5783 2015 | 333.9'2--dc23 | 2015006556 |

Apple Academic Press also publishes its books in a variety of electronic formats. Some content that appears in print may not be available in electronic format. For information about Apple Academic Press products, visit our website at **www.appleacademicpress.com** and the CRC Press website at **www.crcpress.com**

# About the Editor

---

**MUYIWA S. ADARAMOLA, PhD**

**Dr. Muyiwa S. Adaramola** earned his BSc and MSc in Mechanical Engineering from Obafemi Awolowo University, Nigeria, and University of Ibadan, Nigeria, respectively. He received his PhD in Environmental Engineering at the University of Saskatchewan in Saskatoon, Canada. He has worked as lecturer at the Obafemi Awolowo University and as a researcher at the Norwegian University of Science and Technology, Trondheim, Norway. Currently, Dr. Adaramola is a Professor in Renewable Energy at the Norwegian University of Life Sciences, Ås, Norway.

# Contents

# Acknowledgment and How to Cite

The editor and publisher thank each of the authors who contributed to this book. The chapters in this book were previously published in various places in various formats. To cite the work contained in this book and to view the individual permissions, please refer to the citation at the beginning of each chapter. Each chapter was read individually and carefully selected by the editor; the result is a book that provides a nuanced look at the implications of using wind energy. The chapters included are broken into three sections, which describe the following topics:

- Access to water is a serious problem in Africa, as well as in other parts of the world. Chapter 1's research is important, since it indicates that wind energy is a practical and useful resource for water pumping in regions where the water supply is limited.
- Some scientists hypothesize that global warming may produce changed weather patterns that will detracts from wind power's future viability. The data within Chapter 2, however, indicates that on the contrary, wind energy will continue to be a stable energy source throughout the twenty-first century.
- The research in Chapter 3 finds that the expansion of wind energy is an economic boon to most regions. However, the authors do not include the potential environmental costs in their analysis, which may weaken their findings in the minds of some critics of wind energy. Consequently, we address those issues in later chapters.
- The authors of Chapter 4 make creative use of game software to create visual and acoustic simulations of wind farms, offering a valuable decision-making tool for the placement of future wind farm development.
- In Chapter 5, we investigated wind turbines suitable for wind-energy development along Ghana's coast. We found that the economic analysis supports this development, but it will require government assistance, in the form of financial subsidy, infrastructure development, and manpower training, to accelerate its reality.
- The authors of Chapter 6 investigate the many factors that must be considered when planning for offshore wind-energy development. They acknowledge that in real-world practice not every recommendation can be satisfied, and they call for ongoing and dedicated research that will allow the next generation of offshore wind farms to better balance the ecological needs of

sensitive offshore sites, climate-change targets and legislation, and human economic demands. It is essential that we not see these three issues as being in conflict with each other; ongoing research should instead support the notion that these work together.

- Economic concerns will always be foremost in the minds of some segments of the population. In Chapter 7, the authors created a model-based estimate of the millions of dollars of benefit to local economies from wind development, from both short-term and long-term perspectives.

- In Chapter 8, the authors focus on ways to minimize the risk to wildlife from wind energy development. They intersect a 3-D potential power-generation map with an environmental-impact map in order to give decision-makers information on both the power output and environmental risk of a particular turbine location. Additionally, their analysis predicts that areas where bird populations are high could be excluded from development with minimal consequences to the potential of power generation from the remaining locations.

- The authors of Chapter 9 find that environmental impact of wind-energy development cannot be predicted in a generalized way. The geographic context where wind energy is developed plays an important role, meaning that planners must consider factors such as land cover, biodiversity, water resources, and other environmental and aesthetic factors. In consequence, we cannot make sweeping statements about wind energy being appropriate for all regions of the world, since it must be considered region by region.

- There is much talk in the media about the supposedly detrimental affects to human health from wind turbines. The authors of Chapter 10 conduct a comprehensive review of research findings to date and find no evidence to support this. They do find, however, that turbine noise can annoy people, and that annoyance can lead to subjective assumptions of physical harm.

- The research in Chapter 11 indicates that when dealing with public perceptions such as those discussed in the previous chapter, wind-energy development is facilitated by open public dialogue at the local level.

- A current trend is to place larger wind turbines in remote locations. Chapter 12 investigates the need for better monitoring systems for these wind farms, with improved capacity for failure prognostics.

- The author of Chapter 13 confronts one of the fundamental problems with wind energy: the fluctuating nature of wind creates challenges for the power grid. He calls for a dual solution that increases power system flexibility, combined with locally fine-tuned wind-farm responses to grid needs.

# List of Contributors

**Muyiwa S. Adaramola**
Deaprtment of Ecology and Natural Resource Management, Norwegian University of Life Science, Ås, Norway

**Martin Agelin-Chaab**
Department of Automotive, Mechanical and Manufacturing Engineering, University of Ontario Institute of Technology, Oshawa, ON, Canada

**Melissa L. Whitfield Aslund**
Intrinsik Environmental Sciences Inc., Mississauga, ON, Canada

**Zanita Avotniece**
Faculty of Geography and Earth Sciences, University of Latvia, Alberta 10, Riga 1010, Latvia

**Helen Bailey**
Chesapeake Biological Laboratory, University of Maryland Center for Environmental Science, 146 Williams Street, Solomons, MD 20688, USA

**Robert G. Berger**
Intrinsik Environmental Sciences Inc., Mississauga, ON, Canada

**Gil Bohrer**
Department of Civil, Environmental and Geodetic Engineering, The Ohio State University, Columbus, Ohio, United States of America

**Kate L. Brookes**
Marine Scotland Science, 375 Victoria Road, Aberdeen AB11 9DB, UK

**Roger W. Compton**
U.S. Geological Survey, Geosciences and Environmental Change Science Center, Denver Federal Center, Denver, Colorado, United States of America

**Peter S. Curtis**
Department of, Evolution, Ecology, and Organismal Biology, The Ohio State University, Columbus, Ohio, United States of America

**Patrick Devine-Wright**
Department of Geography, College of Life and Environmental Sciences, University of Exeter, Exeter, UK

**Jay E. Diffendorfer**
U.S. Geological Survey, Geosciences and Environmental Change Science Center, Denver Federal Center, Denver, Colorado, United States of America

**Gabriel Ekemb**
Department of Applied Sciences, University of Quebec, Chicoutimi, QC G7H 2B1, Canada and School of Engineering, University of Quebec, Rouyn-Noranda, QC J9X 5E4, Canada

**Mark Geisken**
Weathernews, Inc., Norman, OK 73069, USA

**John Scott Greene**
Department of Geography and Environmental Sustainability, University of Oklahoma, Norman, OK 73019, USA

**Adrienne Grêt-Regamey**
Planning of Landscape and Urban Systems, Swiss Federal Institute of Technology Zurich, Zurich CH-8093, Switzerland

**Ulrike Wissen Hayek**
Planning of Landscape and Urban Systems, Swiss Federal Institute of Technology Zurich, Zurich CH-8093, Switzerland

**Kurt Heutschi**
Empa, Swiss Federal Laboratories for Materials Science and Technology, Duebendorf CH-8600, Switzerland

**Robert L. Jones**
Department of Civil, Environmental and Geodetic Engineering, The Ohio State University, Columbus, Ohio, United States of America

**Loren D. Knopper**
Intrinsik Environmental Sciences Inc., 1790 Courtwood Crescent, Ottawa ON, K2C 2B5, Canada

**Cornelius Kuba**
Mechanical Engieering, Technical University Hamburg, Harburg, Germany

**Lita Lizuma**
Institute of Physical Research and Biomechanics, Artilerijas 40, Riga 1090, Latvia

**Arif S. Malik**
College of Engineering, Sultan Qaboos University, 123 Al-Khod, Oman

**Madeleine Manyoky**
Planning of Landscape and Urban Systems, Swiss Federal Institute of Technology Zurich, Zurich CH-8093, Switzerland

**Lindsay C. McCallum**
Intrinsik Environmental Sciences Inc., Mississauga, ON, Canada

**Mary McDaniel**
Intrinsik Environmental Sciences Inc., Venice, CA, USA

**Christopher A. Ollson**
Intrinsik Environmental Sciences Inc., 6605 Hurontario Street, Suite 500, Mississauga, ON, L5T 0A3, Canada

**Mohand Ouhrouche**
Department of Applied Sciences, University of Quebec, Chicoutimi, QC G7H 2B1, Canada

**Sunday O. Oyedepo**
Mechanical Engineering Department, Covenant University, Ota, Ogun State, 112101, Nigeria

**Samuel S. Paul**
Department of Mechanical and Manufacturing Engineering, University of Manitoba, Winnipeg, Manitoba, R3T 5V6, Canada

**Reto Pieren**
Empa, Swiss Federal Laboratories for Materials Science and Technology, Duebendorf CH-8600, Switzerland

**Sergejs Rupainis**
Rezekne Higher Education Institution, Atbrivosanas aleja 90, Rezekne 4601, Latvia

**Fouad Slaoui-Hasnaoui**
School of Engineering, University of Quebec, Rouyn-Noranda, QC J9X 5E4, Canada

**Kathleen Souweine**
Intrinsik Environmental Sciences Inc., Venice, CA, USA

**Tommy Andy Tameghe**
Department of Applied Sciences, University of Quebec, Chicoutimi, QC G7H 2B1, Canada and School of Engineering, University of Quebec, Rouyn-Noranda, QC J9X 5E4, Canada

**Pierre Tchakoua**
Department of Applied Sciences, University of Quebec, Chicoutimi, QC G7H 2B1, Canada and
School of Engineering, University of Quebec, Rouyn-Noranda, QC J9X 5E4, Canada

**Artis Teilans**
Institute of Physical Research and Biomechanics, Artilerijas 40, Riga 1090, Latvia

**Paul M. Thompson**
Institute of Biological and Environmental Sciences, Lighthouse Field Station, University of Aberdeen,
George Street, Cromarty, Ross-shire IV11 8YJ, UK

**René Wamkeue**
Department of Applied Sciences, University of Quebec, Chicoutimi, QC G7H 2B1, Canada and
School of Engineering, University of Quebec, Rouyn-Noranda, QC J9X 5E4, Canada

**Xiaoming Yuan**
College of Electrical and Electronic Engineering, Huazhong University of Science and Technology,
Wuhan, 430074, China

**Kunpeng Zhu**
Department of Civil, Environmental and Geodetic Engineering, The Ohio State University, Columbus,
Ohio, United States of America

# Introduction

Humans have used wind power for thousands of years, first for sailing the seas and then for pumping water and grinding grain. Windmills may have been developed in the Middle East and Persia more than two thousand years ago. These very early windmills were mostly the vertical-axis type. A thousand years later, the Dutch refined the windmill to pump water off their flooded land. By the Middle Ages, windmills were used across Europe mainly for mechanical tasks such as water pumping, grinding stone, grinding grain, sawing wood, and powering tools. The horizontal-axis windmills were introduced during this period. Eventually, there were more than 10,000 windmills in England alone. Late in the nineteenth century, Americans made windmills more efficient with lighter, faster steel blades (instead of the wooden blades that had been used for centuries). During the last years of the nineteenth century and the early years of the twentieth century, more than six million small windmills were built in the United States alone. They were mostly used for pumped water for livestock and homes. The first large windmill that produced electricity was built in 1888 in the United States (12-kilowatt rated power and diameter of 17 m), while the first modern wind turbine, specifically designed for electricity generation, was constructed in Denmark in 1890. In 1931, the first utility-scale wind turbine system was installed in Russia. The rural electrification in the United States of America and in Europe led to the decline of wind energy conversion systems use in the late 1930 to 1950s.

Today, as the world looks for alternatives to fossil fuels, wind energy offers the promise of at least a partial solution to humanity's energy needs. If, for example, wind turbines produced three billion kWh of electricity, they would displace the energy equivalent of 6.4 million barrels of oil and avoid 1.67 million tons of carbon emissions, as well as sulfur and nitrogen oxide emissions that cause smog and acid rain. Wind power has become the world's fastest growing energy source. It is also one of the most rapidly expanding energy industries. The global wind energy market was $130 billion in 2013 and \$165.5 billion in 2014, with projection that the wind energy market could be worth about \$250 billion by 2020.

Despite wind energy's enormous potential, however, challenges remain to be overcome. The research gathered in this compendium begins with a brief overview of wind energy's potential in a few locations around the world. In the second part of the compendium, the authors investigate the social, economic, and environmental issues that must be considered in connection with wind energy. The book concludes with a brief overview of current trends as well as the challenges that wind energy must face in the future.

*Muyiwa S. Adaramola, Ph.D*

In Chapter 1, Oyedepo and colleagues investigated the wind speed characteristics and energy potential in three selected locations in the southeastern part of Nigeria using wind speed data that span between 24 and 37 years and measured at a height of 10 m. It was shown that the annual mean wind speed at a height of 10 m for Enugu, Owerri and Onitsha are 5.42, 3.36 and 3.59 m/s, respectively, while the annual mean power densities are 96.98, 23.23 and 28.34 W/m$^2$, respectively. It was further shown that the mean annual value of the most probable wind speed are 5.47, 3.72 and 3.50 m/s for Enugu, Owerri and Onitsha, respectively, while the respective annual value of the wind speed carrying maximum energy are 6.48, 4.33 and 3.90 m/s. The performance of selected commercial wind turbine models (with rated power between 50 and 1,000 kW) designed for electricity generation and a windmill (rated power, 0.36 kW) for water pumping located in these sites was examined. The annual energy output and capacity factor for these turbines, as well as the water produced by the windmill, were determined. The minimum required design parameters for a wind turbine to be a viable option for electricity generation in each location are also suggested.

Offshore wind energy development promises to be a significant domestic renewable energy source in Latvia. The reliable prediction of present and future wind resources at offshore sites is crucial for planning and selecting the location for wind farms. The overall goal of Chapter 2, by Lizuma and colleagues, is the assessment of offshore wind power potential in a target territory of the Baltic Sea near the Latvian coast as well as the identification of a trend in the future wind energy potential for the study territory. The regional climate model CLM and High Resolution Limited Area Model (Hirlam) simulations were used to obtain the wind climatology data for the study area. The results indicated that offshore wind energy is promising for expanding the national electricity generation and will continue to be a stable resource for electricity generation in the region over the 21st century.

Many options can be effectively used to meet the future power needs of a country in ways which would be more economically viable, environmentally sound, and socially just. In Chapter 3, Malik and Kuba conducted a least-cost generation expansion planning study to find the economic feasibility of large scale integration of wind farms in the main interconnected transmission system of Oman. The generation expansion planning software used is WASP which is restricted in its ability to model intermittent nature of wind. Therefore, a wind turbine is modeled as a thermal plant with high forced outage rate related to its capacity factor. The result of the study has shown that wind turbines are economically viable option in the overall least-cost generation expansion plan for the Main Interconnected System of Oman.

Public landscape impact assessment of renewable energy installations is crucial for their acceptance. Thus, a sound assessment basis is crucial in the implementation process. For valuing landscape perception, the visual sense is the dominant human sensory component. However, the visual sense provides only partial information about our environment. Especially when it comes to wind farm assessments, noise produced by the rotating turbine blades is another major impact factor. Therefore, an integrated visual and acoustic assessment of wind farm projects is needed to allow lay people to perceive their impact adequately. Chapter 4, by Manyoky and colleagues, presents an approach of linking spatially referenced auralizations to a GIS-based virtual 3D landscape model. The authors demonstrate how to utilize a game engine for 3D visualization of wind parks, using geodata as a modeling basis. In particular, the controlling and recording of specific parameters in the game engine is shown in order to establish a link to the acoustical model. The resulting prototype has high potential to complement conventional tools for an improved public impact assessment of wind farms.

Chapter 5, by Adaramola and colleagues, examines the wind energy potential and the economic viability of using wind turbine for electricity generation in selected locations along the coastal region of Ghana. The two-parameter Weibull probability density function was employed to analyze the wind speed data obtained from the Ghana Energy Commission. The energy output and unit cost of electricity generated from medium size commercial wind turbine models with rated powers ranging from 50 kW to 250 kW were determined. It was found that the wind resource along the coastal region of Ghana can be classified into Class 2 or less wind resource which indicate that this resource in this area is marginally suitable for large scale wind energy development or suitable for small scale applications and be useful as part of hybrid energy system. It was further observed that wind turbine with designed cut-in wind speed of less than 3 m/s and moderate rated wind speed between 9 and 11 m/s is more suitable for wind energy development along the coastal region of Ghana. Based on the selected wind turbine and assumptions used in this study, it was estimated that the unit cost of electricity varied between 0.0695 GH¢/kW h and 0.2817 GH¢/kW h.

Offshore wind power provides a valuable source of renewable energy that can help reduce carbon emissions. Technological advances are allowing higher capacity turbines to be installed and in deeper water, but there is still much that is unknown about the effects on the environment. In Chapter 6, Bailey and colleagues describe the lessons learned based on the recent literature and our experience with assessing impacts of offshore wind developments on marine mammals and seabirds, and make recommendations for future monitoring and assessment as interest in offshore wind energy grows around the world. The four key lessons learned that we discuss are: 1) Identifying the area over which biological effects may occur to inform baseline data collection and determining the connectivity between key populations and proposed wind energy sites, 2) The need to put impacts into a population level context to determine whether they are biologically significant, 3) Measuring responses to wind farm construction and operation to determine disturbance effects and avoidance responses, and 4) Learn from other industries to inform risk assessments and the effectiveness of mitigation measures. As the number and size of offshore wind developments increases, there will be a growing need to consider the population level consequences and cumulative impacts of these activities on marine species. Strategically targeted data collection and modeling aimed at answering questions for the consenting process will also allow regulators to make decisions based on the best available information, and achieve a balance between climate change targets and environmental legislation.

There have been increasing efforts nationally and internationally to promote renewable energy as a response to the awareness of the limited supply of fossil fuels, to meet growing energy demand, and to reduce the harmful environmental impacts of fossil fuel use. To address these efforts, there have been numerous studies to address the impact to local communities. However, these studies have typically focused on either the economic or the social aspects of the wind farm development. Chapter 7, by Greene and Geisken, analyzes the combined and varied socioeconomic impacts as well as the stakeholder perceptions associated with wind power development in Weatherford, Oklahoma. This project uses a mixed-method approach to investigate the impact on a small city when a substantial wind farm is built nearby. This approach consists of three components: a survey, in-depth personal interviews, and economic modeling. The economic modeling is performed to determine both direct and indirect economic impacts. Results from this research show the economic impact on the local community and estimate the number of construction and other types of jobs. In addition, the interviews and surveys illustrate other aspects of the socioeconomic impact and describe overall attitudes of the population to the wind farm development. The study uses a case study and a mixed methods approach to illustrate the socioeconomic impacts of wind farm development. As the world moves increasingly toward green energy, studies like this are important to be able to fully understand impacts on the local community of this type of development.

The location of a wind turbine is critical to its power output, which is strongly affected by the local wind field. Turbine operators typically seek locations with the best wind at the lowest level above ground since turbine height affects installation costs. In many urban applications, such as small-scale turbines owned by local communities or organizations, turbine placement is challenging because of limited available space and because the turbine often must be added without removing existing infrastructure, including buildings and trees. The need to minimize turbine hazard to wildlife compounds the challenge. In Chapter 8, Bohrer and colleagues used an exclusion zone approach for turbine-placement optimization that incorporates spatially detailed maps of wind distribution and wildlife densities with power output predictions for the Ohio State University campus. The authors processed public GIS records and airborne lidar point-cloud data to develop a 3D map of all campus buildings and trees. High resolution large-eddy simulations and long-term wind climatology were combined to provide land-surface-affected 3D wind fields and the corresponding wind-power generation potential. This power prediction map was then combined with bird survey data. The authors' assessment predicts that exclusion of areas where bird numbers are highest will have modest effects on the availability of locations for power generation. The exclusion zone approach allows the incorporation of wildlife hazard in wind turbine siting and power output considerations in complex urban environments even when the quantitative interaction between wildlife behavior and turbine activity is unknown.

Land transformation (ha of surface disturbance/MW) associated with wind facilities shows wide variation in its reported values. In addition, no studies have attempted to explain the variation across facilities. In Chapter 9, Diffendorfer and Compton digitized land transformation at 39 wind facilities using high resolution aerial imagery. The authors then modeled the effects of turbine size, configuration, land cover, and topography on the levels of land transformation at three spatial scales. The scales included strings (turbines with intervening roads only), sites (strings with roads connecting them, buried cables and other infrastructure), and entire facilities (sites and the roads or transmission lines connecting them to existing infrastructure). An information theoretic modeling approach indicated land cover and topography were well-supported variables affecting land transformation, but not turbine size or configuration. Tilled landscapes, despite larger distances between turbines, had lower average land transformation, while facilities in forested landscapes generally had the highest land transformation. At site and string scales, flat topographies had the lowest land transformation, while facilities on mesas had the largest. The results indicate the landscape in which the facilities are placed affects the levels of land transformation associated with wind energy. This creates opportunities for optimizing wind energy production while minimizing land cover change. In addition, the results indicate forecasting the impacts of wind energy on land transformation should include the geographic variables affecting land transformation reported here.

The association between wind turbines and health effects is highly debated. Some argue that reported health effects are related to wind turbine operation [electromagnetic fields (EMF), shadow flicker, audible noise, low-frequency noise, infrasound]. Others suggest that when turbines are sited correctly, effects are more likely attributable to a number of subjective variables that result in an annoyed/stressed state. In Chapter 10, Knopper and colleagues provide a bibliographic-like summary and analysis of the science around this issue specifically in terms of noise (including audible, low-frequency noise, and infrasound), EMF, and shadow flicker. Now there are roughly 60 scientific peer-reviewed articles on this issue. The available scientific evidence suggests that EMF, shadow flicker, low-frequency noise, and infrasound from wind turbines are not likely to affect human health; some studies have found that audible noise from wind turbines can be annoying to some. Annoyance may be associated with some self-reported health effects (e.g., sleep disturbance) especially at sound pressure levels >40 dB(A). Because environmental noise above certain levels is a recognized factor in a number of health issues, siting restrictions have been implemented in many jurisdictions to limit noise exposure. These setbacks should help alleviate annoyance from noise. Subjective variables (attitudes and expectations) are also linked to annoyance and have the potential to facilitate other health complaints via the nocebo effect. Therefore, it is possible that a segment of the population may remain annoyed (or report other health impacts) even when noise limits are enforced. Based on the findings and scientific merit of the available studies, the weight of evidence suggests that when sited properly, wind turbines are not related to adverse health. Stemming from this review, the authors provide a number of recommended best practices for wind turbine development in the context of human health.

In response to the threat of climate change, many governments have set policy goals to rapidly and extensively increase the use of renewable energy in order to lessen reliance upon fossil fuels and reduce emissions

of greenhouse gases. Such policy goals are ambitious, given past controversies over large-scale renewable energy projects, particularly onshore
wind farms, that have occurred in many countries and involved bitter disputes between private developers and local 'NIMBYs' (not in my backyard) protestors. Chapter 11, by Devine-Wright, critically reviews recent
research into how public engagement is conceived and practiced by policy
makers and developers, with a specific focus upon the UK. The review
reveals a distinction between different scales of technology deployment,
with active public engagement only promoted at smaller scales, and a
more passive role promoted at larger scales. This passive role stems from
the influence of widely held NIMBY conceptions that presume the public
to be an 'ever present danger' to development, arising from a deficit in
factual knowledge and a surfeit of emotion, to be marginalized through
streamlined planning processes and one-way engagement mechanisms. It
is concluded that NIMBYism is a destructive, self-fulfilling way of thinking that risks undermining the fragile, qualified social consent that exists
to increase renewable energy use. Breaking the cycle of NIMBYism requires new ways of thinking and practicing public engagement that better connect national policy making with local places directly affected by
specific projects. Such a step would match the radical ambitions of rapid
increases in renewable energy use with a process of change more likely to
facilitate its achievement.

As the demand for wind energy continues to grow at exponential rates,
reducing operation and maintenance (OM) costs and improving reliability
have become top priorities in wind turbine (WT) maintenance strategies.
In addition to the development of more highly evolved WT designs intended to improve availability, the application of reliable and cost-effective condition-monitoring (CM) techniques offers an efficient approach
to achieve this goal. Chapter 12, by Tchakoua and colleagues, provides
a general review and classification of wind turbine condition monitoring
(WTCM) methods and techniques with a focus on trends and future challenges. After highlighting the relevant CM, diagnosis, and maintenance
analysis, this work outlines the relationship between these concepts and
related theories, and examines new trends and future challenges in the
WTCM industry. Interesting insights from this research are used to point
out strengths and weaknesses in today's WTCM industry and define re-

search priorities needed for the industry to meet the challenges in wind industry technological evolution and market growth.

Wind power has been developing rapidly in major countries in the past 10 years. The distinct static and dynamic characteristics of output power compared with conventional generations pose significant challenges on power system adequacy and stability and constraints on the penetration level of wind power in power systems. Based on the uniqueness of wind power versus conventional generations, in Chapter 13, Yuan discusses its implications on power system adequacy and stability and propose basic solutions for facilitating large-scale integrations of wind power into the power system.

# PART I

# WIND ENERGY'S POTENTIAL ASSESSMENT IN SELECTED LOCATIONS

# Analysis of Wind Speed Data and Wind Energy Potential in Three Selected Locations in South-East Nigeria

SUNDAY O. OYEDEPO, MUYIWA S. ADARAMOLA, AND SAMUEL S. PAUL

## 1.1 BACKGROUND

The quest to reduce environmental impacts of conventional energy resources and, more importantly, to meet the growing energy demand of the global population had motivated considerable research attention in a wide range of environmental and engineering application of renewable form of energy. It is recognized that wind energy, as a renewable energy source, has stood out as the most valuable and promising choice. Wind energy by nature is clean, abundant, affordable, inexhaustible and environmentally preferable. Due to its many advantages, wind energy has also become the fastest growing renewable source of energy in both developed and developing countries. For example, wind energy is widely used to produce electricity in countries like Denmark, Spain, Germany, USA, China and India. Interestingly, the global cumulative installed capacity of wind power had

*Analysis of Wind Speed Data and Wind Energy Potential in Three Selected Locations in South-East Nigeria. © Oyedepo SO, Adaramola MS, and Paul SS. International Journal of Energy and Environmental Engineering **3**,7 (2012), doi:10.1186/2251-6832-3-7. Licensed under Creative Commons Attribution 2.0 Generic License, http://creativecommons.org/licenses/by/2.0.*

increased sharply from 6,100 MW in 1996 to about 237,669 MW in 2011 [1]. In Africa, for example, Egypt, Morocco and Tunisia are the leading countries with installed capacities of 550, 291 and 114 MW, respectively, at the end of 2011 [1].

The increasing energy demand, the rapidly depleting fossil fuel reserves and the environmental problems associated with the use of fossil fuel have necessitated the development of alternative energy sources like wind energy for electricity generation in Nigeria. It is reported that the electricity production in Nigeria as of the end of 2010 is less than 4,000 MW due to fluctuations in the availability and maintenance of production sources, leading to a shortfall in supply [2]. However, analyses of available wind data for selected cities have confirmed a high prospect of wind energy resources in Nigeria. Several studies on renewable sources of energy have also been performed. A detailed review and discussion of these studies can be found in [2-6] and are not repeated here. Worthy of mention here from these studies, however, is that the effective utilization of wind energy at a typical location requires sound knowledge of the wind characteristics and accurate wind data analysis. For example, the choice of wind turbine design must be based on the average wind velocity at a selected wind turbine installation site [7]. Prior studies have also shown that the wind flow patterns are influenced by terrains, vegetation and water bodies.

**TABLE 1:** The geographical location of the selected stations

| Station | Latitude (N) | Longitude (E) | Altitude (m) | Measurement period |
|---------|--------------|---------------|--------------|--------------------|
| Enugu   | 6° 26′       | 7° 29′        | 304.70       | 1971 to 2007       |
| Owerri  | 5° 29′       | 7° 02′        | 186.05       | 1977 to 2002       |
| Onitsha | 6° 10′       | 6° 47′        | 63.14        | 1978 to 2003       |

*N, North; E, East.*

Although several studies have been performed to investigate the characteristics and pattern of wind speed across Nigeria, less attention has been given to sites in the south-east region. According to [8,9], the few reported studies on wind speed in this part of the country were limited to wind speed distributions, while less attention was paid to the wind energy potential evaluation. The focus of this study is, therefore, to evaluate the wind energy potential in three selected locations (Enugu, Owerri and Onitsha) in the south-east region and to assess the performance of selected small- to medium-size commercial wind turbines. It is the authors' view that this information will be helpful to the government and any organization in making an informed decision with regard to investment in wind energy resource in this part of Nigeria.

## 1.2 METHODS

The wind data used in this study were obtained from the Nigerian Meteorological Agency, Oshodi, Lagos. The geographical coordinates of the meteorological stations where the wind speed data were captured at a height of 10 m by a cup-generator anemometer are given in Table 1. There are many sources of measurement uncertainty in cup-anemometer measurements. The guidelines and steps necessary to minimize these errors are outlined in Manwell et al. [10]. Following the methodologies proposed and explained in the ISO guide [11] to the expression of uncertainty in measurement, the uncertainty in the mean velocities at 95% confidence level was determined to be ±2%. Monthly wind data that span between 24 and 37 years were obtained for Enugu, Owerri and Onitsha. The recorded wind speeds were computed as the mean of the speed for each month. It should be noted that using monthly wind speed has some limitations such as loosing extremely low or high wind speeds within the month as well as inability to observe diurnal variations in the wind speed. However, using monthly mean wind speed, which is mostly available for most locations, can be used to study the seasonal changes in wind speed and facilitates wind data analysis.

### 1.2.1 FREQUENCY DISTRIBUTION
### AND SITE WIND SPEED PARAMETERS

Several mathematical models such as normal and lognormal have been used for wind data analysis. Prior studies have also shown that statistical methods such as the Weibull and Rayleigh distribution models can equally be used [12]. According to [13-15], the two-parameter Weibull probability distribution function is the most appropriate, accepted and recommended distribution function for wind speed data analysis. This is because it gives a better fit for measured monthly probability density distributions than other statistical functions [12,15]. In addition, the Weibull parameters at known height can be used to estimate wind parameters at another height [13]. Therefore, the two-parameter Weibull probability density function was used in this study. In Weibull distribution, the variation in wind velocity is characterized by two parameter functions: the probability density function and the cumulative distribution. The probability density function f(V) indicates the probability of the wind at a given velocity V, while the corresponding cumulative distribution function of the velocity V gives the probability that the wind velocity is equal to or lower than V, or within a given wind speed range. The Weibull probability density function is given as, e.g., [12,16]:

$$f(V) = \left(\frac{k}{c}\right)\left(\frac{V}{c}\right)^{k-1} \exp\left[-\left(\frac{V}{c}\right)^{k}\right] \tag{1}$$

where $f(V) =$ the probability of observing wind speed (V), k = dimensionless Weibull parameter and c = the Weibull scale parameter (in meter per second). The scale factor could be related to the mean wind speed through the shape factor, which determines the uniformity of the wind speed in a given site. The cumulative distribution F(V) is the integral of the probability density function, and it is expressed as, e.g., [12,16]:

$$F(V) = 1 - e^{-\left(\frac{V}{c}\right)^{k}} \tag{2}$$

The monthly and annual values of Weibull parameters were calculated using standard deviation method. This method is useful where only the mean wind speed and standard deviation are available. In addition, it gives better results than graphical method and has relatively simple expressions when compared with other methods [13,17,18]. Moreover, it is unlike most of the other methods that may require more detailed wind data (which, in some cases, are not readily available) for the determination of the Weibull distribution shape and scale parameters. The shape and scale factors are thus computed from Equations 3 and 4 given by [13,19]:

$$k = \left(\frac{\sigma}{V_m}\right)^{-1.086} \tag{3}$$

$$c = \frac{V_m}{\Gamma\left(1 + \frac{1}{k}\right)} \tag{4}$$

where $\sigma$ is the standard deviation, $V_m$ is the mean wind speed (in meter per second) and $\Gamma(x)$ is the gamma function, which is defined as [17,19]:

$$\Gamma(x) = \int_0^\infty t^{x-1} e^{-t} dt \tag{5}$$

Alternatively, scale factor can be determined from the following expressions given by [20]:

$$c = \frac{V_m k^{2.6674}}{0.184 + 0.816 k^{2.73855}} \tag{6}$$

Equation 6 is used in this study to estimate the monthly and annual scale factors.

In addition to the mean wind speed, the other two significant wind speeds for wind energy estimation are the most probable wind speed ($V_F$) and the wind speed carrying maximum energy ($V_E$). They can be expressed respectively as [12,21]:

$$V_F = c \left( \frac{k-1}{k} \right)^{1/k} \tag{7}$$

$$V_E = c \left( \frac{k+2}{k} \right)^{1/k} \tag{8}$$

The most probable wind speed corresponds to the peak of the probability density function, while the wind speed carrying maximum energy can be used to estimate the wind turbine design or rated wind speed. Prior studies have shown that wind turbine system operates most efficiently at its rated wind speed. Therefore, it is required that the rated wind speed and the wind speed carrying maximum energy should be as close as possible [16].

## 1.2.2 EXTRAPOLATION OF WIND SPEED AT DIFFERENT HUB HEIGHT

In most cases, the available wind data are measured at a height different from the wind turbine hub height. It is noted that it is the wind speed at the hub height that is of interest for wind power application; therefore, the available wind speeds are adjusted to the wind turbine hub height using the following power law expression, e.g., [12]:

$$\frac{V}{V_o} = \left( \frac{h}{h_o} \right)^{\alpha} \tag{9}$$

where V is the wind speed at the hub height, $hV_o$ is wind speed at the original height $h_o$, and $\alpha$ is the surface roughness coefficient and is assumed to be 0.143 (or 1/7) in most cases. The surface roughness coefficient $\alpha$ can be determined from the following expression [22]:

$$\alpha = [0.37 - 0.088\ln(V_0)]/\left[1 - 0.088\ln\left(\frac{h_o}{10}\right)\right] \tag{10}$$

Alternatively, the Weibull probability density function can be used to obtain the extrapolated values of wind speed at different heights. Since the boundary layer development and the effect of the ground are non-linear with respect to wind speed, the scale factor c and form factor k of the Weibull distribution will change as a function of height by the following expressions [13]:

$$c(h) = c_o\left(\frac{h}{h_o}\right)^n \tag{11}$$

$$k(h) = k_o\left[1 - 0.088\ln\left(\frac{h_o}{10}\right)\right]/\left[1 - 0.088\ln\left(\frac{h}{10}\right)\right] \tag{12}$$

where $c_0$ and $k_0$ are the scale factor and shape parameter, respectively, at the measurement height ho. The exponent n is defined as:

$$n = [0.37 - 0.088\ln(c_0)]/\left[1 - 0.088\ln\left(\frac{h}{10}\right)\right] \tag{13}$$

## 1.2.3 MEAN WIND POWER DENSITY AND ENERGY DENSITY

The mean wind power density can be estimated by using the following equation:

$$P_D = \frac{P(V)}{A} = \frac{1}{2}\rho V_m^3 \tag{14}$$

where $P(V)$=the wind power (in watt), $P_D$=the wind power density (watt per square meter), $\rho$=the air density at the site (assumed to be 1.225 kg/m³ in this study) and $A$=the swept area of the rotor blades (in square meter). Both the mean wind speed and power density are generally used to classify the wind energy resource (e.g., Pacific Northwest Laboratory (PNL) wind power classification scheme, Illica et al. [23]). However, the wind power density (wind power per unit area) based on the Weibull probability density function can be calculated using the following equation [24]:

$$P_D = \frac{P(V)}{A} = \frac{1}{2}\rho c^3 \Gamma \left(1 + \frac{3}{k}\right) \tag{15}$$

The mean energy density ($E_D$) over a period of time T is the product of the mean power density and the time T, and it is expressed as:

$$E_D = \frac{1}{2}\rho c^3 \Gamma \left(1 + \frac{3}{k}\right) T \tag{16}$$

## 1.2.4 WIND TURBINE ENERGY OUTPUT AND CAPACITY FACTOR

A wind energy conversion system can operate at its maximum efficiency only if it is designed for a particular site because the rated power and cut-in and cut-off wind speeds must be defined based on the site wind characteristics [12]. It is essential that these parameters are selected so that energy output from the conversion system is maximized. The performance of a wind turbine installed in a given site can be examined by the amount of mean power output over a period of time ($P_{e,ave}$) and the conversion efficiency or capacity factor of the turbine. The capacity factor $C_f$ is defined as the ratio of the mean power output to the rated electrical power ($P_{eR}$) of the wind turbine [12,20].

The mean power output $P_{e,ave}$ and capacity factor $C_f$ of a wind turbine can be estimated using the following expressions based on Weibull distribution function [12]:

$$P_{e,ave} = P_{eR} \left( \frac{e - \left(\frac{vc}{c}\right)^k - e - \left(\frac{vr}{c}\right)^k}{\left(\frac{vr}{c}\right)^k - \left(\frac{vc}{c}\right)^k} - e - \left(\frac{vf}{c}\right)^k \right) \tag{17}$$

$$C_f = \frac{P_{e,ave}}{P_{eR}} \tag{18}$$

where $v_c$ $v_r$ and $v_f$ are the cut-in wind speed, rated wind speed and cut-off wind speed, respectively. For an investment in wind power to be cost effective, it is suggested that the capacity factor should be greater than 0.25 [25].

## 1.2.5  WIND-DRIVEN ROTODYNAMIC PUMPS

There are three types of wind-powered pumping systems. They are the mechanical-piston pump, the mechanical-air lift (rotodynamic) pump and the electrical pump. In general, the volume of water produced by rotodynamic and electrical pumps are considered to be more than that of a piston pump at the same wind speed regime. This is because there is a better match between the rotodynamic and electrical pumps and the wind rotor than for a piston pump [26,27]. In this study, the performance of a rotodynamic pump is simulated. For a rotodynamic pump driven by a wind turbine with a given cut-in wind speed and a cut-out wind speed, the water produced over a period of time T can be determined from:

$$Q = T \int_{V_i}^{V_0} Q(V)f(V)dV$$

$$(19)$$

where Q(V) is the discharge of the pump at any wind speed, and it is given by [28]:

$$Q(V) = \frac{1}{8}C_P\eta VD \left(\frac{\rho_a}{\rho_w}\right) \left(\frac{V_d^2}{gH}\right) \left(\frac{G\lambda_D}{N_{PD}}\right)$$

$$(20)$$

where $V_d$ is design wind speed, G is the gear ratio, $N_{PD}$ is the speed of the pump at design condition, D is the wind turbine rotor diameter and $\lambda_D$ is the design tip speed ratio of the wind turbine. For a water-pumping application, a tip speed ratio between 1 and 3 is recommended [10]. By substituting Equation 20 into Equation 19 and assuming Rayleigh probability density function f(V), the total water produced over a given time is expressed as [26]:

$$Q = \frac{\pi}{16Vm^2}C_{P_D}\eta_{PD}TD \left(\frac{\rho_a}{\rho_w}\right) \left(\frac{V_d^2}{gH}\right) \left(\frac{G\lambda_D}{N_{PD}}\right) \int_{V_i}^{V_0} V^2 \exp - \left(\frac{\pi}{4}\left(\frac{V}{V_m}\right)^2\right) dV$$

$$(21)$$

## 1.3 RESULTS AND DISCUSSION

### 1.3.1 WIND SPEED FREQUENCY DISTRIBUTION

The annual probability density frequency and cumulative distributions of wind speed for the three locations obtained using the Weibull distribution function are shown in Figure 1. The probability density function is used to illustrate the fraction of time for which given wind speed possibly prevails at a location. As expected, the peak of the density function frequencies of all the sites skewed towards the higher values of mean wind speed (Figure 1a). It should be remarked that the peak of the probability density function curve indicates the most frequent velocity. It can be observed from Figure 1a that the most frequent wind speed expected in Enugu, Owerri and Onitsha are about 5.5, 3.5 and 3.5 m/s, respectively. It can be further observed that Enugu has the highest spread of wind speed toward high wind speed among the locations.

The cumulative probability distributions of the wind speed at all the study locations (Figure 1b) show a similar trend. The cumulative distribution function can be used for estimating the time for which wind speed is within a certain speed interval. For wind speeds greater or equal to 2.5 m/s cut-in wind speed, Enugu, Owerri and Onitsha have frequencies of about 96.9%, 86.5% and 86.9%, respectively, while the same locations respectively have frequencies of about 88.4%, 44.7% and 55.3% for wind speed of 3.5 m/s cut-in wind speed. According to Ojosu and Salawu [3], if a wind turbine system with a design cut-in wind speed of 2.2 m/s is used in these sites for wind energy resource for electricity generation, all the sites will have frequencies of more than 92%.

## 1.3.2 MEAN WIND SPEED AND MEAN POWER DENSITY

The monthly variation of the mean wind speed characteristics ($V_m$, $V_F$ and $V_E$), mean power density and mean energy density as well as the annual values of these parameters at a height of 10 m are presented in Tables 2, 3 and 4.

The monthly mean wind speed varies between 4.13 m/s in November and 6.30 m/s in March for Enugu site (Table 2). The monthly mean power density varies between 43.15 W/m² in November and 153.16 W/m² in March. Therefore, based on PNL wind power classification scheme [23], the monthly mean power density mostly falls into class 1 (PD≤100) except in January, February, April and July, when it falls into class 2 (100<PD≤150), and in March, when it falls into class 3 (150<PD≤200). However, the annual mean power density for this site is 96.98 W/m² (class

1). For Owerri (Table 3), the minimum and maximum values of the monthly mean wind speeds are 2.72 and 3.70 m/s, respectively, while the annual mean wind speed for this site is 3.36 m/s. The monthly mean power density varies between 11.66 W/m² in November and 31.02 W/m² in January. The monthly mean power density falls into class 1 wind resource category (PD≤100) in all the months, and the annual mean power density for this site is 23.23 W/m² (class 1). In the case of Onitsha (Table 4), the minimum and maximum values of the monthly mean wind speeds are 3.01 m/s (in November) and 4.23 m/s (in March), respectively. The monthly mean power density varies between 16.70 W/m² in November and 46.36 W/m² in March. The monthly mean power density falls into class 1 wind resource category (PD≤100) in all the months, and the annual mean power density for this site is 28.34 W/m² (class 1). Detailed information about these sites' wind speed characteristics (mean wind speed, most probable wind speed (VF) and the wind speed carrying maximum energy (VE)) and mean power density are illustrated in Tables 2, 3 and 4.

**TABLE 2:** Characteristic speeds and mean power density in Enugu at a height of 10 m

|        | $V_m$(m/s) | k    | c(m/s) | $V_F$(m/s) | $V_E$(m/s) | $P_D$(W/m²) | $E_D$(kWh/m²) |
|--------|-----------|------|--------|-----------|-----------|------------|--------------|
| Jan    | 5.62      | 3.49 | 6.25   | 5.68      | 7.12      | 108.72     | 78.279       |
| Feb    | 5.67      | 4.86 | 6.19   | 5.90      | 6.64      | 111.65     | 80.387       |
| Mar    | 6.30      | 5.75 | 6.81   | 6.58      | 7.17      | 153.15     | 110.271      |
| Apr    | 6.22      | 5.11 | 6.77   | 6.49      | 7.22      | 147.39     | 106.123      |
| May    | 5.35      | 4.44 | 5.87   | 5.54      | 6.39      | 93.79      | 67.530       |
| Jun    | 5.21      | 4.82 | 5.69   | 5.42      | 6.11      | 86.62      | 62.367       |
| Jul    | 5.48      | 5.45 | 5.94   | 5.72      | 6.29      | 100.80     | 72.574       |
| Aug    | 5.44      | 3.93 | 6.01   | 5.58      | 6.68      | 98.61      | 70.996       |
| Sept   | 4.85      | 4.79 | 5.30   | 5.04      | 5.70      | 69.88      | 50.311       |
| Oct    | 4.56      | 5.24 | 4.96   | 4.76      | 5.27      | 58.08      | 41.815       |
| Nov    | 4.13      | 3.75 | 4.58   | 4.22      | 5.13      | 43.15      | 31.066       |
| Dec    | 4.95      | 3.34 | 5.52   | 4.96      | 6.36      | 74.29      | 53.488       |
| Annual | 5.42      | 4.05 | 5.87   | 5.47      | 6.48      | 96.98      | 717.619      |

$V_m$, mean wind speed; k, dimensionless Weibull shape parameter; c, Weibull scale parameter; $V_F$, most probable wind speed; $V_E$, wind speed carrying maximum energy; $P_D$, wind power density; $E_D$, mean energy density.

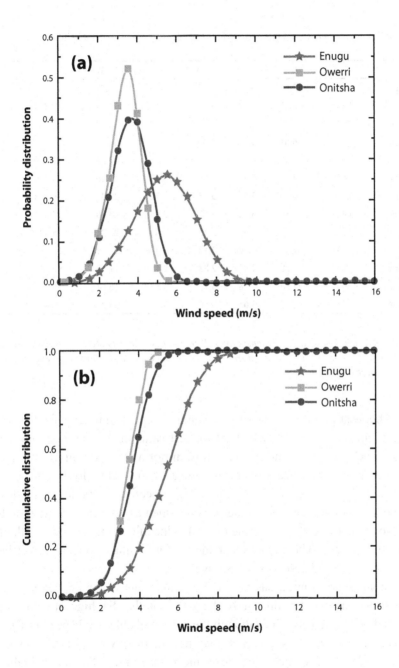

**FIGURE 1:** Annual wind speed distribution. (a) Probability density function and (b) cumulative density function.

**TABLE 3:** Characteristic speeds and mean power density in Owerri at a height of 10 m

|        | $V_m$(m/s) | k    | c(m/s) | $V_F$(m/s) | $V_E$(m/s) | $P_D$(W/m²) | $E_D$(kWh/m²) |
|--------|------------|------|--------|------------|------------|-------------|----------------|
| Jan    | 3.52       | 2.61 | 3.96   | 3.29       | 4.92       | 26.71       | 19.234         |
| Feb    | 3.54       | 8.11 | 3.74   | 3.68       | 3.84       | 27.17       | 19.564         |
| Mar    | 3.64       | 7.00 | 3.87   | 3.79       | 4.02       | 29.54       | 21.269         |
| Apr    | 3.70       | 6.61 | 3.96   | 3.87       | 4.12       | 31.02       | 22.338         |
| May    | 3.45       | 8.48 | 3.63   | 3.58       | 3.72       | 25.15       | 18.109         |
| Jun    | 3.55       | 8.46 | 3.74   | 3.68       | 3.83       | 27.40       | 19.730         |
| Jul    | 3.32       | 6.30 | 3.56   | 3.47       | 3.72       | 22.41       | 16.138         |
| Aug    | 3.42       | 5.30 | 3.72   | 3.57       | 3.95       | 24.50       | 17.641         |
| Sept   | 3.31       | 7.01 | 3.53   | 3.45       | 3.66       | 22.21       | 15.993         |
| Oct    | 3.04       | 5.41 | 3.30   | 3.17       | 3.49       | 17.21       | 12.390         |
| Nov    | 2.67       | 6.28 | 2.97   | 2.88       | 3.10       | 11.66       | 8.394          |
| Dec    | 3.00       | 4.80 | 3.28   | 3.12       | 3.53       | 16.54       | 11.907         |
| Annual | 3.36       | 5.10 | 3.65   | 3.50       | 3.90       | 23.23       | 203.530        |

$V_m$, mean wind speed; k, dimensionless Weibull shape parameter; c, Weibull scale parameter; $V_P$ most probable wind speed; $V_E$, wind speed carrying maximum energy; $P_D$, wind power density; $E_D$, mean energy density.

The least monthly value of the Weibull shape parameter k for Owerri is 2.61 in January and reached the highest value of 8.48 in the month of May. Therefore, the wind speed is most uniform in Owerri in May, while it is least uniform in December. The annual shape factors for Enugu, Owerri and Onitsha are 4.05, 5.10 and 4.27, respectively. The least monthly value of Weibull scale parameter c is obtained as 2.97 m/s in the month of November in Owerri, and the highest value of 6.81 m/s in the month of March in Enugu. The annual shape factors for Enugu, Owerri and Onitsha are 5.87, 3.65 and 3.96 m/s, respectively.

In summary, Enugu has the highest annual mean wind speed among the sites considered in this study. Also, this site has the highest values of annual power density. Even though the most probable wind speed (VF) is a statistical characteristic, which may not be directly connected to wind energy [21], it does not necessarily mean that Enugu has much higher wind potentials than the other locations considered. However, as mentioned earlier, the efficiency of a wind turbine is closely related to these parameters, especially VE, which should be as close as possible to the design or rated wind speed of the system. Therefore, the proposed wind turbine, if installed in Enugu, would likely produce more power than other locations. Moreover, it can be considered as the best site for wind energy development in southern Nigeria (based on the three sites considered in this study). Furthermore, the monthly mean wind speeds in south Nigeria ranges from 2.72 to 6.30 m/s. The monthly mean power density varies between 11.66 and 153.15 W/m², while the annual mean power density is in the range of 23.23 to 96.98 W/m². It can be inferred from this analysis that the wind resource in this part of Nigeria can be classified mostly into class 2 or less category. Furthermore, the annual mean energy density varies between 203.53 and 717.62 kWh/m².

**TABLE 4:** Characteristic speeds and mean power density in Onitsha at a height of 10 m

|       | $V_m$(m/s) | k | c(m/s) | $V_F$(m/s) | $V_E$(m/s) | $P_D$(W/m²) | $E_D$(kWh/m²) |
|-------|-----------|------|--------|-----------|-----------|------------|--------------|
| Jan   | 3.59 | 4.91 | 3.92 | 3.74 | 4.20 | 28.34 | 20.404 |
| Feb   | 3.73 | 4.11 | 4.11 | 3.84 | 4.53 | 31.79 | 22.886 |
| Mar   | 4.23 | 4.86 | 4.62 | 4.41 | 4.96 | 46.36 | 33.378 |
| Apr   | 4.07 | 5.00 | 4.44 | 4.24 | 4.75 | 41.29 | 29.732 |
| May   | 3.76 | 4.71 | 4.11 | 3.91 | 4.43 | 32.56 | 23.442 |
| Jun   | 3.72 | 4.55 | 4.08 | 3.86 | 4.42 | 31.53 | 22.702 |
| Jul   | 3.58 | 4.51 | 3.92 | 3.71 | 4.26 | 28.10 | 20.234 |
| Aug   | 3.61 | 4.77 | 3.95 | 3.76 | 4.25 | 28.82 | 20.747 |
| Sept  | 3.52 | 4.74 | 3.85 | 3.66 | 4.15 | 26.71 | 19.234 |
| Oct   | 3.28 | 4.23 | 3.60 | 3.37 | 3.94 | 21.61 | 15.562 |
| Nov   | 3.01 | 4.78 | 3.29 | 3.13 | 3.54 | 16.70 | 12.026 |
| Dec   | 3.05 | 3.52 | 3.39 | 3.08 | 3.85 | 17.38 | 12.512 |
| Annual| 3.59 | 4.27 | 3.96 | 3.72 | 4.33 | 28.34 | 248.252 |

$V_m$, mean wind speed; k, dimensionless Weibull shape parameter; c, Weibull scale parameter; $V_P$ most probable wind speed; $V_E$ wind speed carrying maximum energy; $P_D$ mean wind power density; $E_D$ mean energy density.

Even though the wind resource in these locations falls into class 2 or less, which is considered as marginally or unsuitable for wind power development, the wind power can be used for water pumping and small-scale electricity generation, providing intermittent power requirements for a variety of purposes that need low-energy capacity, slow-running high-torque wind turbines with multi-blade, e.g., [3,4,6,29]. For a modern wind turbine, the cut-in wind speed required by it to start generating electricity is generally between 3 to 5 m/s. Depending on the size of the turbine, the peak power output can be attained when the wind speed (rated wind speed) is in the range of 10 to 15 m/s [6]. For water pumping, wind turbine can be operated at a lower wind speed; however, they can function effectively when the wind speed is more than 3 m/s. Based on the required quantity of water, a site with a mean wind speed around 2.0 m/s can be considered for wind-powered pump application [30]. Similarly, depending on the end use of the generated power, it can be concluded that these locations may be suitable for utilization of wind energy.

### 1.3.3 PERFORMANCE OF SELECTED WIND TURBINES

Seven small- to medium-size commercial wind turbine models with rated power range from 50 to 1,000 kW [31-33] were selected to simulate their performance at Enugu, Owerri and Onitsha. These are P15-50, P19-100, P50-500 and P62-1000 models (Polaris America LLC, Lakewood, NJ, USA ); WES30 model (Wind Energy Solutions BV, The Netherlands); WWD-1-60 model (Winwind, Espoo, Finland) and BONUS 1000–54 (Siemens AG, Erlangen, Germany). The selected wind turbine models and their characteristic properties are given in Table 5. For each location, the annual energy output and capacity factor based on Weibull distribution function parameters at their respective hub height are determined using Equations 17 and 18, respectively.

The performance of the selected wind turbine models at all the locations is presented in Figure 2. The figure clearly reveals that irrespective of the wind turbine model, Enugu seems to be the best site for wind power development for electricity generation. This is expected because, when compared with other sites, Enugu has the highest annual mean wind speed and highest value of VE at the hub height for each turbine. The annual energy output for Enugu ranges from about 113 MWh/year using PO-LARIS 19–100 model to 2,444 MWh using WWD-1-60 model. Among the 1,000-kW model turbines, the WWD-1-60 model produced the highest power output, closely followed by the POLARIS 62–1000 (2,431 MWh/year), while BONUS 1000–54 produced the least power (849 MWh/year). This observed trend is related to the hub height and rotor diameter (which are lowest for the BONUS model), the design or rated wind speed (highest for the BONUS model, 14 m/s) and the cut-in wind speed (highest for the WWD-1-60 model, 3.6 m/s). Even though the hub height of the WWD-1-60 model is higher than that of the POLARIS 62 model, the POLARIS 62 model still produced almost the same amount of power (2,431 MWh/year) due to its lower cut-in wind speed and rated wind speed. Furthermore, it is observed that the energy output from the POLARIS 15–50 model is slightly more than the power output from the POLARIS 19–100 model. This is because both models have the same hub height, and the POLARIS 19–100 model has higher rated wind speed than the POLARIS 15–50 model.

**TABLE 5:** Characteristics of the selected wind turbines

| | PO-LARIS P15-50 | PO-LARIS P19-100 | WES30 | PO-LARIS P50-500 | POLAR-IS P62-1000 | WWD-1-60 | BO-NUS-1000-54 |
|---|---|---|---|---|---|---|---|
| Rated power (kW) | 50 | 100 | 250 | 500 | 1,000 | 1,000 | 1,000 |
| Hub height (m) | 30 | 30 | 36 | 50 | 60 | 70 | 45 |
| Rotor diameter (m) | 15.2 | 19.1 | 30 | 50 | 62 | 60 | 54 |
| Cut-in wind speed (m/s) | 2.5 | 2.5 | 2.7 | 2.5 | 2.5 | 3.6 | 3 |
| Rated wind speed (m/s) | 10 | 12 | 12.5 | 12 | 12 | 12.5 | 14 |
| Cut-out wind speed (m/s) | 25 | 25 | 25 | 25 | 25 | 25 | 25 |

*[12,31-33].*

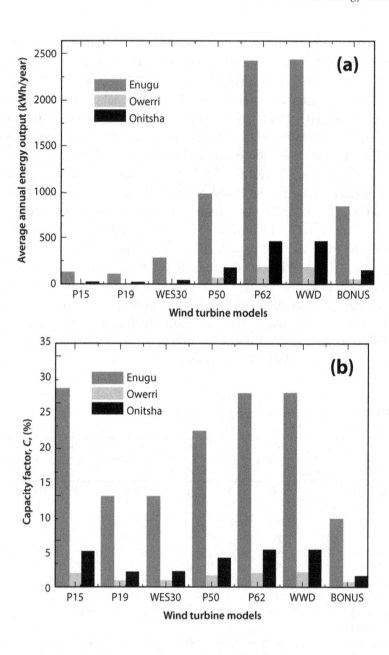

**FIGURE 2:** The performance of the selected wind turbine models for all the locations. (a) Average annual energy output and (b) capacity factor.

The minimum annual energy outputs of 6.05 and 18.83 MWh/year are observed for Owerri and Onitsha, respectively, using the P19-100 model. While the maximum annual energy outputs are 178.58 and 461.32 MWh/year, respectively, for Owerri and Onitsha with the WWD-1-60 model. For each of these sites, the power generated by each wind turbine follows the same trend observed in Enugu. Regardless of the location, the WWD-1-60 wind turbine model produced the highest quantity of annual energy output. For example, if 1,000-kW turbines are to be operated at the same hub height, the POLARIS 62–1000 will likely perform better than WWD-1-60 and BONUS-1000-54 because of its low cut-in wind speed and rated wind speed as well as its bigger rotor diameter compared with other models.

The cost effectiveness of a wind turbine can be roughly estimated by the capacity factor of the turbine. This factor is a useful parameter for both consumer and manufacturer of the wind turbine system [24]. The WWD-1-60 model has the highest value among the models considered for all the sites. The Cf values for this model are 27.90%, 2.04% and 5.27% for Enugu, Owerri and Onitsha, respectively. The Cf values for Enugu for PO-LARIS 15–50, POLARIS 62–1000 and WWD-1-60 are 28.30%, 27.75% and 27.90%, respectively. These values are greater than the suggested recommended value before an investment can be considered worthwhile. Hence, these wind turbines or similar turbine model will be good for electricity generation in Enugu. In Owerri, however, the capacity factor for other wind turbine models ranges from 0.49% for the BONUS-1000-54 model to 2.04% for the WWD-1-60 model. Similarly, the minimum and maximum capacity factors at Onitsha are 1.62% for the BONUS-1000-54 model and 5.27% for the WWD-1-60 model, respectively. Therefore, these two sites may be considered for wind energy development for small-scale applications such as water pumping (see 'Wind-powered pumps performance'). It should be noted that the cost of generating electricity using wind turbine is inversely proportional to the capacity factor. The higher the capacity factor (or higher wind speed regime), the lower the cost of generated electricity, assuming that all factors remain the same (see, e.g., Paul et al. [34]).

Based on the annual energy output and the capacity factor, the PO-LARIS 62–1000 and WWD-1-60 models or wind turbines with similar designed characteristics will be best suited for electricity generation at

Enugu and small-scale application in other locations. However, by rede-
signing the selected wind turbine models to operate at lower cut-in wind
speed (especially the WWD-1-60 and BONUS-100-54 models), lower
rated wind speed (especially BONUS-100-54) and higher hub height compared with their current design parameters (cut-in and rated wind speeds
and hub height), both the annual energy output and capacity factor could
significantly be improved. For instance, if POLARIS 62–1000 is to be operated at a hub height of 70 m and rated wind speed of 10 m/s, the capacity
factor for Enugu, Owerri and Onitsha will be 59.17%, 8.20% and 16.99%,
respectively. However, increasing the hub height may increase the overall
capital cost of the wind turbines. But this is compensated for by increased
in capacity factor and, hence, the energy output from the wind turbines.

In order to meet the minimum recommended capacity factor (25%)
for electricity generation, the following design parameters are suggested: wind turbine model with a minimum hub height of 55 m, cut-in wind
speed of less than 3.5 m/s, rated wind speed of around 12 m/s and cut-out
wind speed of 25 m/s are recommended for Enugu; for Owerri, wind turbine with a minimum height of 75 m, cut-in wind speed of less than 3.5
m/s, rated wind speed of around 8.5 m/s and cut-out wind speed of 20 m/s
are recommended; while a system with a minimum hub height of 65 m,
cut-in wind speed of less than 3.5 m/s, rated wind speed of around 9 m/s
and cut-out wind speed of 20 m/s are recommended for Onitsha.

## 1.3.4 WIND-POWERED PUMPS PERFORMANCE

In this section, the performance of wind-powered pumps assumed to be
located in each of the locations considered in this study is presented and
discussed. For the performance assessment, a wind turbine model with a
rated power of about 0.36 kW is chosen. The characteristic properties of
this wind turbine model and the specifications of the pumps are given in
Table 6. The wind turbine characteristics are similar to the FT-2.5 wind
turbine model (produced in China, according to [35]) [35,36] for water
pumping. The hub height of the wind turbine is assumed to be 15 m, which
is within the hub height range for common commercial wind turbines for
water-pumping application [30]. The pump head of 10 m is used for the

analysis. This pump head was chosen because it is within the range of the water table level of 3 to 20 m in this part of Nigeria [37] and within the range of the design lifting height of 5 to 10 m for the FT-2.5 windmill [35]. The monthly mean wind speeds at hub height at each selected site are computed using Equations 9 and 10. The quantity of water pumped per month is computed using Equation 23. This equation was solved using Wind Energy Resource Analysis software developed by Mathew [25]. The software is based on Rayleigh distribution function and requires the mean wind speed, wind turbine and pump parameters as inputs.

**TABLE 6:** Wind turbine parameters and rotodynamic pump specifications

| Parameters/specifications | Value |
|---|---|
| Wind turbine | |
| Diameter (m) | 2.5 |
| Rated power (W) | 360 |
| Rated speed (m/s) | 7 |
| Cut-in speed (m/s) | 2.5 |
| Cut-out speed (m/s) | 10 |
| Design speed ratio | 2 |
| Design power coefficient | 0.35 |
| Pump | |
| Efficiency (pump and transmission) | 0.6 |
| Gear ratio | 10 |

The comparison between water produced per month by rotodynamic pump at the three locations is presented in Table 7. As expected from the wind speed frequency distribution, the amount of water output from the pump at Enugu is highest among the three locations. In Enugu, the monthly water output varies between about 3,136 and 3,512 m³, and the average volume discharge is about 3,442 m³/month. In Owerri, where the pump has the least performance among the three sites, the monthly water output varies between 1,936 and 2,844 m³ with an average discharge volume of about 2,530 m³/month. In the case of Onitsha, the monthly water output varies between about 2,206 and 3,194 m³, and the average volume discharge is about 2,730 m³/month. Therefore, all the sites have strong potential for wind energy development for water pumping. In fact, wind resource in these sites could also be used for electricity generation, as previously shown.

The total numbers of habitants that can be served by water discharged from these sites are also shown in Table 7. Based on water usage of 36 L/capita/day in Nigeria as of 2006 [38], the water output at Enugu can serve between 2,900 and 3,250 habitants depending on the month. The average number of people that can be served per month is estimated to be around 3,190. However, if the estimate is based on the minimum recommended water usage of 50 L/capita/day [38], the water produced can only serve about 2,290 habitants per month on average. In Owerri, the water produced can serve about 2,340 and 1,690 habitants per month on average based on 36 and 50 L/capita/day, respectively, while the total numbers of habitants that can be served by the water produced in Onitsha are 2,530 and 1,820/month on average based on water usage of 36 and 50 L/capita/day, respectively. Therefore, for small rural communities scattered across the southeastern part of Nigeria where access to good water and unreliable supply of water is a regular problem, wind resource development for water pumping will be a good option. For a larger community, the performance of a pump can be increased if a wind turbine with higher rated power (or diameter) is used instead of the small size used in this study. In addition, two or more wind turbines can be installed in these sites in order to increase the quantity of water produced.

## 1.4 CONCLUSIONS

In this study, the wind speed and wind energy potential in selected three locations in the southeastern part of Nigeria were investigated. In addition, the performance of selected commercial wind turbine models designed for both electricity generation and water pumping located in these sites was investigated. The findings from this study can be summarized as follows:

**TABLE 7:** Monthly water produced and the number of habitants that can be served per month

| Month | Enugu | | | | Owerri | | | | Onitsha | | | |
|---|---|---|---|---|---|---|---|---|---|---|---|---|
| | $V_m$ (m/s) | Q (m³) | Habitants 50 L/day | Habitants 36 L/day | $V_m$ (m/s) | Q (m³) | Habitants 50 L/day | Habitants 36 L/day | $V_m$ (m/s) | Q (m³) | Habitants 50 L/day | Habitants 36 L/day |
| Jan | 6.16 | 3,511.51 | 2,341 | 3,251 | 3.92 | 2,687.47 | 1,792 | 2,488 | 4.00 | 2,752.48 | 1,835 | 2,549 |
| Feb | 6.21 | 3,510.02 | 2,340 | 3,250 | 3.95 | 2,710.74 | 1,807 | 2,510 | 4.15 | 2,859.98 | 1,907 | 2,648 |
| Mar | 6.88 | 3,436.48 | 2,291 | 3,182 | 4.05 | 2,787.89 | 1,859 | 2,581 | 4.69 | 3,194.29 | 2,130 | 2,958 |
| Apr | 6.79 | 3,450.62 | 2,300 | 3,195 | 4.12 | 2,843.95 | 1,896 | 2,633 | 4.52 | 3,100.75 | 2,067 | 2,871 |
| May | 5.87 | 3,507.62 | 2,338 | 3,248 | 3.85 | 2,633.46 | 1,756 | 2,438 | 4.18 | 2,885.82 | 1,924 | 2,672 |
| Jun | 5.72 | 3,495.80 | 2,331 | 3,237 | 3.96 | 2,721.07 | 1,814 | 2,520 | 4.14 | 2,857.08 | 1,905 | 2,645 |
| Jul | 6.01 | 3,512.36 | 2,342 | 3,252 | 3.71 | 2,510.67 | 1,674 | 2,325 | 3.98 | 2,738.42 | 1,826 | 2,536 |
| Aug | 5.97 | 3,511.42 | 2,341 | 3,251 | 3.82 | 2,605.90 | 1,737 | 2,413 | 4.03 | 2,770.28 | 1,847 | 2,565 |
| Sept | 5.34 | 3,432.44 | 2,288 | 3,178 | 3.70 | 2,502.78 | 1,669 | 2,317 | 3.93 | 2,692.31 | 1,795 | 2,493 |
| Oct | 5.04 | 3,343.72 | 2,229 | 3,096 | 3.41 | 2,232.75 | 1,489 | 2,067 | 3.65 | 2,460.31 | 1,640 | 2,278 |
| Nov | 4.58 | 3,136.10 | 2,091 | 2,904 | 3.11 | 1,936.35 | 1,291 | 1,793 | 3.38 | 2,205.90 | 1,471 | 2,043 |
| Dec | 5.45 | 3,456.25 | 2,304 | 3,200 | 3.37 | 2,192.40 | 1,462 | 2,030 | 3.42 | 2,240.39 | 1,494 | 2,074 |

*Hub height is assumed as 15 m. Vm, the mean wind speed; Q, water produced at a given time.*

1. The annual mean wind speeds for Enugu, Owerri and Onitsha are 5.42, 3.36 and 3.59 m/s, respectively. The annual values of the wind speed carrying maximum energy for these locations are respectively 6.48, 4.33 and 3.90 m/s.

2. The mean annual value of Weibull shape parameter k is between 4.05 and 5.10, while the annual value of scale parameter c is between 3.96 and 5.87 m/s.

3. The annual mean power density for Enugu, Owerri and Onitsha are 96.98, 23.23 and 28.34 $W/m^2$, respectively. Therefore, based on the wind data used in this study, the wind energy resource in south-east Nigeria may generally be classified into class 1. However, based on monthly mean power density, the wind resource may fall into higher class category in some cases.

4. Based on the capacity factor, the POLARIS 15–50, POLARIS 62–1000 and WWD-1-60 models or wind turbines with similar designed characteristics will be best suited for electricity generation in Enugu. However, in order to meet the minimum recommended capacity factor (25%) for electricity generation, wind turbine models with cut-in wind speed of less than 3.5 m/s minimum and hub height of 55, 75 and 65 m, as well as rated wind speed of about 12, 8.5 and 9 m/s, respectively, are recommended for Enugu, Owerri and Onitsha.

5. Using a 0.36-kW wind turbine, the average monthly water produced by a rotodynamic pump assumed to be installed in Enugu, Owerri and Onitsha is determined as 3,442, 2,530 and 2,730 $m^3$, respectively. The quantity of water can serve about 2,290, 1,690 and 1,820 habitants in respective locations.

## REFERENCES

1. Global Wind Energy Council: Global wind report: annual market updates. http://www.gwec.net/ webcite (2011). Accessed 30 April 2012

2. Fagbenle, RO, Katende, J, Ajayi, OO, Okeniyi, JO: Assessment of wind energy potential of two sites in North-East. Nigeria. Renew. Energy. 36, 1277–1293 (2011)

3. Ojosu, JO, Salawu, RI: A survey of wind energy potential in Nigeria. Solar and Wind Technology. 7, 155–67 (1990).

4. Adekoya, LO, Adewale, AA: Wind energy potential of Nigeria. Renew. Energy. 2, 35–39 (1992).
5. Fadare, DA: The application of artificial neural networks to mapping of wind speed profile for energy application in Nigeria. Applied Energy. 87, 934–942 (2010).
6. Adaramola, MS, Oyewola, OM: On wind speed pattern and energy potential in Nigeria. Energy Policy. 39, 2501–2506 (2011).
7. Marcius-Kaitis, M, Katnals, V, Karaliauskas, A: Wind power usage and prediction prospects in Lithuania. Renew. Sustain. Energy Rev.. 12, 265–277 (2008).
8. Anyanwa, EE, Iwuagwu, CJ: Wind characteristics and energy potentials for Owerri Nigeria. Renew. Energy. 6, 125–128 (1995).
9. Asiegbu, AD, Iwuoha, GS: Studies of wind resources in Umudike, south east Nigeria-an assessment of economic viability. J. Eng. Appl. Sci.. 2, 1539–1541 (2007)
10. Manwell, JF, McGowan, JG, Rogers, AL: Wind Energy Explained: Theory, Design and Application, Wiley, Wiltshire (2010)
11. International Standards Organisation: Guide to the Expression of Uncertainty in Measurement. (1992)
12. Akpinar, EK, Akpinar, S: An assessment on seasonal analysis of wind energy characteristics and wind turbine characteristics. Energy Conversion and Management. 46, 1848–67 (2005).
13. Justus, CG, Hargraves, WR, Mikhail, A, Graber, D: Methods for estimating wind speed frequency distributions. J. Appl. Meteorol.. 17, 350–353 (1978).
14. Akdag, SA, Dinler, A: A new method to estimate Weibull parameters for wind energy applications. Energy Conversion and Management. 50, 1761–1766 (2009).
15. Akdag, SA, Bagiorgas, HS, Mihalakakou, G: Use of two-component Weibull mixtures in the analysis of wind speed in the Eastern Mediterranean. Applied Energy. 87, 2566–2573 (2010).
16. Mathew, S, Pandey, KP, Kumar, AV: Analysis of wind regimes for energy estimation. Renew. Energy. 25, 381–399 (2002).
17. Jowder, FAL: Wind power analysis and site matching of wind turbine generators in Kingdom of Bahrain. Applied Energy. 86, 538–545 (2009).
18. Kwon, SD: Uncertainty analysis of wind energy potential assessment. Applied Energy. 87, 856–865 (2010).
19. Ouammi, A, Dagdougui, H, Sacile, R, Mimet, A: Monthly and seasonal assessment of wind energy characteristics at four monitored locations in Liguria region (Italy). Renew. Sustain. Energy Rev.. 14, 1959–1968 (2010).
20. Balouktsis, A, Chassapis, D, Karapantsios, TD: A nomogram method for estimating the energy produced by wind turbine generators. Solar Energy. 72, 251–259 (2002).
21. Bagiorgas, HS, Assimakopoulos, MN, Theoharopoulos, D, Matthopoulos, D, Mihalakakou, GK: Electricity generation using wind energy conversion systems in the area of Western Greece. Energy Conversion and Management. 48, 1640–1655 (2007).
22. Ucar, A, Balo, F: Evaluation of wind energy potential and electricity generation at six locations in Turkey. Applied Energy. 86, 1864–1872 (2009).
23. Ilinca, A, McCarthy, E, Chaumel, J-L, Retiveau, J-L: Wind potential assessment of Quebec Province. Renew. Energy. 28, 1881–1897 (2003).

24. Celik, AN: A statistical analysis of wind power density based on the Weibull and Rayleigh models at the southern region of Turkey. Renew. Energy. 29, 593–604 (2004).

25. Mathew, S: Wind Energy: Fundamentals. Resource Analysis and Economics, Springer, Heidelberg (2006)

26. Mathew, S, Pandey, KP: Modelling the integrated output of wind-driven roto-dynamic pumps. Renew. Energy. 28, 1143–1155 (2003).

27. Hau, E, Renouard, R: Wind turbines: Fundamentals, technologies, applications, economics, Springer, Heidelberg (2006)

28. Mathew, S, Pandey, KP: Modelling the integrated output of mechanical wind pumps. J. Solar Energy Engineering. 122, 203–206 (2000).

29. Fagbenle, RL, Karayiannis, TG: On the wind energy resource of Nigeria. Intern. J. Energy Res.. 18, 493–508 (1994).

30. Meel, J, Smulders, P: Wind Pumping: A Handbook (World Bank Technical Paper No. 101), World Bank, Washington (1989)

31. Polaris: http://www.polarisamerica.com/ webcite (2011). Accessed 04 April 2011

32. Wind Energy Solutions: http://www.wes30.com/ webcite (2011). Accessed 04 April 2011

33. WinWinD: http://www.winwind.com/Documents/Press%20Kit/wwd1.pdf webcite (2011). Accessed 04 April 2011

34. Paul, SS, Oyedepo, SO, Adaramola, MS: Economic assessment of water pumping systems using wind energy conversion systems in the southern part of Nigeria. Energy Exploration and Exploitation. 30, 1–18 (2012).

35. Shi, J, Shen, D, Wei, J: The development of wind pumping technology in China. Biomass. 20, 13–23 (1989).

36. Ltd, Bobs Harries Engineering: Kijito wind pumps – Pumping water with wind: Information package, Brochure, Bob Harries Engineering Limited (2009).

37. Adelana, SMA, Olashinde, PI, Bale, RB: An overview of the geology and hydrogeology of Nigeria. In: Adelana S, MacDonald A (eds.) Applied Groundwater Studies in Africa, pp. 171–197. Taylor and Francis, London (2008)

38. Programme, United Nations Development: Human Development Report, United Nations Development Programme, New York (2006)

# CHAPTER 2

# Assessment of the Present and Future Offshore Wind Power Potential: A Case Study in a Target Territory of the Baltic Sea Near the Latvian Coast

LITA LIZUMA, ZANITA AVOTNIECE, SERGEJS RUPAINIS, AND ARTIS TEILANS

## 2.1 INTRODUCTION

The exploitation of renewable energy sources can help the European Union meet many of its environmental and energy policy goals, including its obligation to reduce greenhouse gases under the Kyoto Protocol and the aim of securing its energy supply [1].

The current situation regarding wind energy production in Latvia is unsatisfactory. According to the data from Latvenergo AS, approximately 1535 MW (76%) of the total electricity production in Latvia, is generated by hydroelectric power plants, while about 474 MW (23%) is generated at thermal power plants and from fossil fuels, but only around one percent is generated by wind power. Offshore wind energy development promises to be a significant domestic renewable energy source for Latvia.

_Assessment of the Present and Future Offshore Wind Power Potential: A Case Study in a Target Territory of the Baltic Sea Near the Latvian Coast. © Lizuma L, Avotniece Z, Rupainis S, and Teilans A. The Scientific World Journal **2013** (2013). http://dx.doi.org/10.1155/2013/126428. Licensed under Creative Commons Attribution 3.0 Unported License, http://creativecommons.org/licenses/by/3.0/._

The first stage of exploiting the wind energy is the evaluation of wind resources at wind farm sites which means a site-specific evaluation of wind climatology and vertical profiles of wind, as well as the assessment of historical and potential future changes in the wind climate. Suitable long-term wind observation data are needed for this purpose. Reliable prediction of the wind resources as well as site conditions at offshore sites is crucial for project planning and selecting suitable locations. There is a lack of such data sets in the focus area of this study. In situ measurements of wind and environmental conditions are available only for the coastal areas.

There are different ways of estimating the wind resources at a site: interviews of people with local knowledge to identify areas with high and/or low wind speed, measurements only, the measure-correlate-predict method, using global databases, wind atlas methodology, site data-base modelling, and mesoscale and microscale modelling [2, 3]. Recently offshore wind energy mapping has benefited from the advantages provided by the remote sensing data [4–6].

The infrastructure of large offshore installation is typically more expensive than onshore counterparts, relying on somewhat even longer depreciation time. For understanding how the global climate change might influence the magnitude of wind resources as well as the operation and maintenance conditions of wind turbines the General Circulation (or Global Climate) Models (GCMs) and regional climate downscaling (i.e., use of GCM outputs in higher spatial and temporal resolution models) are the primary tools for developing such projections [3]. The method of using a Regional Climate Models (RCMs) to assess the impact of climate change on the wind energy resource has been used before [7, 8].

The overall goal of this paper is the assessment of offshore wind power potential in a target territory of the Baltic Sea near the Latvian coast as well as the identification of a trend (if any) in the future wind energy potential for the study territory.

## 2.2 DATASETS AND CALCULATION METHODS

The assessment of the wind climate was performed for the study region domain extending from 56.03N 20.2E to 57.22N 21.33E (Figure 1). The previous studies have shown that this territory is with a high wind energy generation potential [1]. Several wind turbines are planned to be mounted here in the near future.

## 2.2.1 MODELS DATASETS

In this stage of investigation we focus on the assessment of the long-term present and future conditions. The method of regional climate modelling was used. The regional climate model and high resolution operational model simulations were used to obtain the meteorological and climatological data for the study area.

The RCM simulations used in this research were obtained from ENSEMBLES EU FP6 project (http://www.ensembles-eu.org/) [9–11]. The RCM simulations from ENSEMBLES project have been extensively evaluated [12, 13] and ETHZ-CLM_SCN_HadCM3Q0_25 km (CLM) were chosen for obtaining the climate data. CLM is the climate version of the "Local Model," the operational weather forecast model of the Consortium for Small scale Modelling (COSMO). The consortium was formed by a number of meteorological services including the German Meteorological Service and Meteo Swiss with the goal to "develop, improve, and maintain a nonhydrostatic limited-area atmospheric model, to be used both for operational and for research applications" (http://www.cosmo-model.org/). The results used in this study are from two runs of CLM model, one with the boundary conditions from ERA-40 for the control period 1981–2010 and one with a single set of lateral boundary conditions from ECHAM5 for two future periods 2021–2050 and 2071–2100 applied for the IPCC scenario A1B. The A1B scenario equals to the future world of very rapid economic growth, global population that peaks in the mid-century and declines thereafter, and the rapid introduction of new and more efficient technologies with balance across all sources [14]. The sensitivity of the projections to differences between emissions scenarios was found to be small prior to about 2040, but it increases substantially in the second half of the 21st century [15]. As the emission scenario uncertainty is still relatively small for the period preceding 2050, we focused on climate change under the A1B scenario.

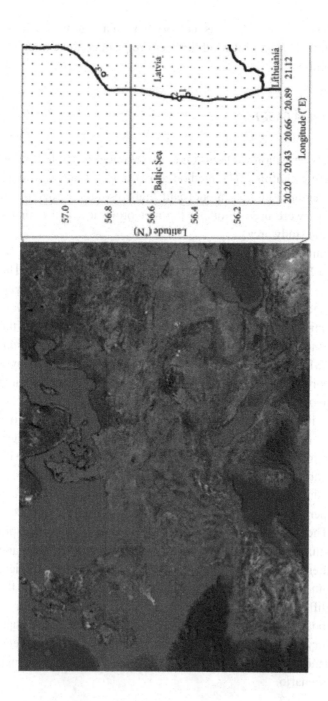

**FIGURE 1:** The study area with sites of In situ observation stations and grid point locations. 1: weather station Liepaja 1; 2: weather station Liepaja 2; 3: weather station Pavilosta.

The meteorological data were obtained also from the Baltic Operational Oceanography System forecast model (DMI-Hirlam) and the EU Project MyOcean FiMAr information system [16]. The data from Hirlam model covers the period 2005–2010.

The modelled data of wind speed, wind direction, air temperature, and atmospheric pressure were used. All meteorological parameters are presented at 10-m, 80 m, 90 m, and 100 m height.

The atmospheric components of CLM were interpolated using Kriging interpolation method from grid cells with horizontal 25 km × 25 km to resolution of 5 × 5 km. Grid cells points are shown in Figure 1. For data comparison purpose the same grid points were used for Hirlam model data. The details of the model data are given in Table 1.

**TABLE 1:** Model simulations used.

| Model (abbreviation) | Global model | Temporal resolution | Periods | Data |
|---|---|---|---|---|
| Regional climate model ETHZ-CLM_SCN_Had-CM3Q0_25 km (CLM) | HadCM3 | Daily (00:00 UTC) | 1981–2010 | Wind speed |
| | | | 2021–2050 | Wind direction |
| | | | 2071–2100 | Air temperature |
| | | | | Atmospheric pressure |
| | | | | Atmospheric model |
| DMI-Hirlam (Hirlam) | ECMWF | Hourly | 2005–2010 | Wind speed |
| | | | | Wind direction |
| | | | | Air temperature |
| | | | | Atmospheric pressure |

## 2.2.2 IN SITU OBSERVATION DATA

Direct measurements from meteorological stations within the study area were obtained from three costal stations (Figure 1). All the stations measured winds at 10 m height. The meteorological stations Liepaja 1 and Pavilosta covered 30 years of observations. Meteorological station Liepaja 2 is located very close to the sea coast and covered 7 years of observations. The details of the in situ observation data are given in Table 2. It should

be noted that taking into account the local site conditions the wind measurement data are representative for the actual measurement site. When
comparing the model output with the observed station data, the model data
were interpolated onto the latitude-longitude station location. The estimation and elimination of local effect associated with site conditions (obstacles and roughness length) as well as complexity of coastal winds were
carried out. The long-terms average monthly and annual wind speed data
of stations Liepaja 1 and Pavilosta were adjusted to roughness length 0.03.

**TABLE 2:** In situ onshore weather stations used in the analysis.

| Station | Distance from the closest coastline, km | Estimated roughness length | Meteorological parameter | Observation period | Temporal resolution |
|---------|---------|---------|---------|---------|---------|
| Liepaja 1 | 1.5 | 0.35 | Wind speed | 1981–2010 | 3-hour |
| | | | Wind direction | | (1981–2003) |
| | | | Air temperature | | Hourly |
| | | | Atmospheric pressure | | (2004–2010) |
| Liepaja 2 | 0.05 | 0.03 | Wind speed | 2004–2010 | Hourly |
| | | | Wind direction | | |
| Pavilosta | 0.3 | 0.74 | Wind speed | 1981–2010 | 3-hour |
| | | | Wind direction | | (1981–2003) |
| | | | Air temperature | | Hourly |
| | | | Atmospheric pressure | | (2004–2010) |

## 2.2.3 METEOROLOGICAL SATELLITE DATA

The data from CLM and Hirlam models were compared with data from
meteorological satellites. The wind speed data at 10 m were obtained from
EUMETSAT established Satellite Application Facility (SAF) on Climate
Monitoring (CM SAF, http://www.cmsaf.eu/) dataset named the "Hamburg Ocean-Atmosphere Parameters and Fluxes from Satellite Data"
(HOAPS) [17]. Satellite data on the wind speed at 10 m height cover a
time period 1998–2005; however, the data are available only at a distance

from the coast, depending on the resolution with which the data were obtained. In some places, this may imply that observations are available only over deep water. Even in these areas, wind statistics from satellites can provide the basic starting point to decide where to focus attention for wind energy purposes. In this study available satellite observation data points which were the closest to the territory of interest were used for the model data evaluation.

## 2.2.4 CALCULATION METHODS

The mean wind speed and the mean cubed wind speed were calculated at each grid point per month, per season, and per day. The mean absolute difference metric (MAD) and the root mean square difference metric (RMSD) were used to compare the CLM and Hirlam simulations with the in situ and satellite observation data.

The "projected percentage change" metric was used to give a measure of expected climate change by comparing the future climate projections with the control simulation. It is defined as follows:

$$D_i = 100 * \frac{(F_i - P_i)}{P_i} \tag{1}$$

where i is the grid point, and $P_i$ and $F_i$ are the wind data for the past and future model runs, respectively.

The wind power density (WPD) was calculated as follows:

$$\text{WPM}\left(\frac{W}{m^2}\right) = 0.5 * \frac{1}{n} * \sum (\rho j * V j^3) \tag{2}$$

where V is the wind speed reading and $\rho$ is air density and n is the number of readings.

Since the air temperature and pressure data are available, the air density was calculated as follows:

$$\rho = \frac{P}{RT} (kg/m^3)$$

(3)

where P = air pressure (in units of Pascals), R = the specific gas constant ($287\,J\,kg^{-1}\,Kelvin^{-1}$), and T = air temperature in degrees Kelvin (deg. C + 273).

## 2.3 RESULTS AND DISCUSSION

### 2.3.1 COMPARISON OF MODEL AND IN SITU DATA

Evaluation of wind speed and direction, air temperature, and atmospheric pressure data from models and in situ data was performed by comparing the mean annual, seasonal, and daily data from three meteorological in situ observation stations (Figure 1) and CLM and Hirlam data at 10 m height.

A good agreement was found between the long-term modelled and observed wind speed data sets in terms of long-term month-to-month variability. The CLM overestimates the long-term annual wind speed by 0.1 m/s (Table 3(a)). On seasonal base the MAD does not exceed 0.3°C. The simulated wind shows generally the same directional pattern as the measurements. The bias of air temperature for the long-term seasonal means does not exceed 0.5°C (Table 3(a)). The observed daily data have a higher correlation coefficient (0.84–0.93 for wind speed; 0.95–0.99 for air temperature; and 0.98-0.99 for air pressure) with the Hirlam data (Table 3(b)) than with CLM data (0.55–0.76 for wind speed; 0.65–0.77 for air temperature; and 0.70–0.89 for air pressure). Comparing the Hirlam simulations and Liepaja 1 and Liepaja 2 in situ daily observations (period 2005–2010) the difference statistics are MAD = 0.3 m/s and 0.6 m/s; RMSD = 1.0 m/s and 0.8 m/s.

It was found that in general the long-term average wind climate predictions of CLM and Hirlam are in good agreement with the in situ measurements. Hirlam model also shows very good agreement with the observation data on daily basis.

**TABLE 3:** Results of comparison of global models and in situ data.

*(a) MAD (m/s) between the simulated and observed long-term seasonal and annual wind speed (m/s) and air temperature (°C) data (1981–2010)*

|  | Winter | Spring | Summer | Autumn | Year |
|---|---|---|---|---|---|
| Wind speed |  |  |  |  |  |
| L1-CLM | −0.1 | 0.3 | 0.0 | 0.1 | 0.1 |
| P-CLM | 0.1 | 0.2 | 0.1 | 0.1 | 0.1 |
| L2-CLM (2004–2010) | 0.1 | 0.1 | 0.1 | 0.1 | 0.1 |
| Air temperature |  |  |  |  |  |
| L1-CLM | 0.2 | 0.1 | 0.3 | 0.2 | 0.4 |
| P-CLM | 0.4 | 0.3 | 0.3 | 0.4 | 0.5 |

*L1 is observation station Liepaja 1; L2 is observation station Liepaja 2; P is observation station Pavilosta. Data have been computed for the four standard climatological seasons (e.g., winter is identified by December, January, February, and so on).*

*(b) Correlation coefficients between daily wind speed, air temperature, and atmospheric pressure (2004–2010)*

| Observation station-model | Winter | Spring | Summer | Autumn | Year |
|---|---|---|---|---|---|
| Wind speed |  |  |  |  |  |
| L1-Hirlam | 0.93 | 0.92 | 0.93 | 0.93 | 0.91 |
| L2-Hirlam | 0.84 | 0.90 | 0.92 | 0.93 | 0.86 |
| P-Hirlam | 0.86 | 0.91 | 0.92 | 0.92 | 0.89 |
| Air temperature |  |  |  |  |  |
| L1-Hirlam | 0.98 | 0.98 | 0.95 | 0.98 | 0.95 |
| P-Hirlam | 0.97 | 0.99 | 0.95 | 0.98 | 0.93 |
| Air pressure |  |  |  |  |  |
| L1-Hirlam | 0.99 | 0.99 | 0.99 | 0.99 | 0.99 |
| P-Hirlam | 0.98 | 0.99 | 0.98 | 0.99 | 0.99 |

*L1 is observation station Liepaja 1; L2 is observation station Liepaja 2; P is observation station Pavilosta. Data have been computed for the four standard climatological seasons (e.g., winter is identified by December, January, February, and so on).*

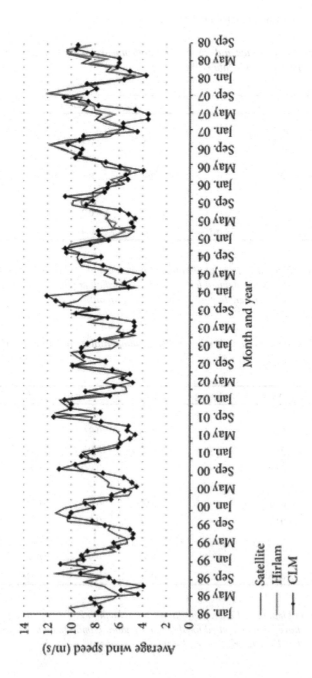

**FIGURE 2:** Time series of monthly average wind speed calculated from CLM, Hirlam, and HOAPS data at grid point 56.03N 20.12E.

## 2.3.2 COMPARISON OF GLOBAL MODEL AND SATELLITE DATA

Monthly average wind speed at 10 m derived from the HOAPS data base was compared with simulation results of the CLM and Hirlam models. During the comparison it was found that a mean absolute difference of 0.3 m/s to 1.0 m/s is evident for Hirlam and CLM data sets, respectively. The bias is mostly positive, and the satellite winds tend to be larger than those of the atmospheric models. Figure 2 illustrates a high agreement between the monthly average wind speed at 10 m height with the correlation coefficients 0.95 (Hirlam) and 0.66 (CLM). The fluctuations of model data around the satellite readings suggest that the small disagreement is due to the differences in temporal resolutions between these data sets. In Figure 2 the time series of monthly average wind speed from modelled and satellite data are illustrated in the proximity of (56.03N 20.12E), located on our grid.

In general the results of this study have shown that both model data are complementary to observations at meteorological stations and satellite readings and can be used for meaningful mapping and assessment of the long-term wind climate and wind power resources for the target offshore area. The Hirlam model has shown a better agreement with the observation data and could be used for an accurate wind regime and wind resources assessment. CLM model appears to be a valuable data source upon which one could base long-term monthly or yearly wind resources estimates and forecasts.

## 2.3.3 RESULTS OF PRESENT WIND CLIMATE ASSESSMENT

The annual wind power potential has a high economical impact for wind parks. The annual power production varies with the wind speed, air temperature and pressure, and from that by the climatology of these elements. The standard procedure to determine the climatology of wind and other meteorological parameters has been to look at 30-year averages (WMO-sanctioned norm), typically the period 1961–1990. In this work we chose to update the 30-year period to a more recent one, encompassing the later part of our data set, 1981–2010. Consequently, the last decade with relative strong anthropogenic impact on the climate was included. However our data from Hirlam model which has shown the better agreement with observation data covers only a 6-year period 2005–2010. In order to assess the differences in the wind climatology between two time periods 1981–2010 and 2005–2010, we evaluated the CLM data for these two time periods. Test shifting the reference period to 2005–2010 showed negligible differences in the results. The average annual wind speed differs from 0.0 to 0.2 m/s and calculated annual wind power density differs from 0 to 3%, respectively, in 492 grid points.

Figure 3 shows the average annual wind speed at 100 m for the time period 2005–2010 and annual average wind power density. A clear gradient in both variables can be observed from the open sea to coastal areas.

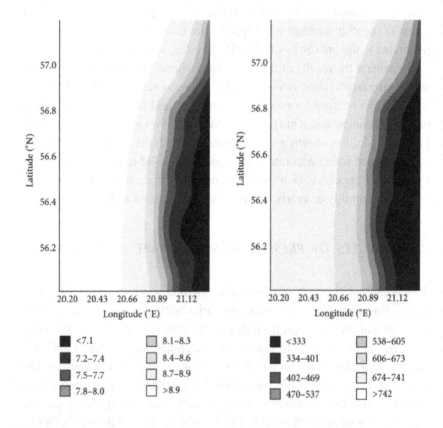

| ■ <7.1 | □ 8.1–8.3 |
| ■ 7.2–7.4 | □ 8.4–8.6 |
| ■ 7.5–7.7 | □ 8.7–8.9 |
| ■ 7.8–8.0 | □ >8.9 |

| ■ <333 | □ 538–605 |
| ■ 334–401 | □ 606–673 |
| ■ 402–469 | □ 674–741 |
| ■ 470–537 | □ >742 |

**FIGURE 3:** (a) Annual average wind speed at 100 m height for the period 2004–2010 based on the Hirlam model data. (b) Average wind power density at 100 m.

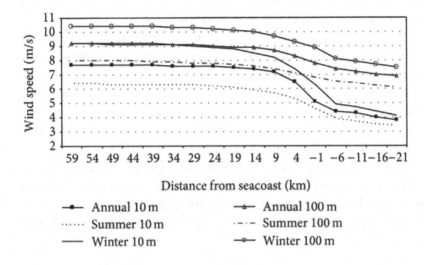

**FIGURE 4:** Mean annual and seasonal (winter-DJF, summer-JJA) wind speed at 10 and 100 m height (2004–2010). The location of the transect is indicated in Figure 1.

The created maps are based on the atmospheric model data that provide a better estimate of the offshore wind resources than the previously available evaluation using the in situ observation data from the coastal stations. In terrestrial territories the wind speeds vary dramatically in the horizontal direction a few kilometres or even hundreds of meters over the sea; however, there are strong local effects near the coasts, but at the open sea much larger areas have fairly uniform wind speeds. Wide areas of similar wind speed can be seen for the region covered in maps. For the most of the sea territory the annual mean wind speed is 6–7.5 m/s at 10 m and 7.5–8.5 m/s at 100 m (Figure 3). The monthly average wind speed in the sea territory varies from 7.5–9.5 m/s in winter to 4-5 m/s in summer at 10 m height and from 9–11 m/s in winter to 6–8 m/s in summer at 100 m height. By contrast, the terrestrial sites have substantially lower wind speed and power density. For all locations and all heights the winter averages are about 1.8 : 1 greater than those of summer. The wind power density increases rapidly by crossing the coast headed seawards. Open sea sites result in about 2.5 to 1.5 greater wind power density than terrestrial locations. The wind power density in all teritory is typically 300–700 W/m². Offshore, the resources are predicted to be more than 500 W/m².

A horizontal transect shown in Figure 4 visualizes the variations in mean wind speed from the coastline to the open part of the sea and illustrates the seasonal variability of the wind speed. The figure shows a significant increase in the wind speed up to 10 km from the coastline to the open sea areas and a slight decrease with the increasing distance from the coast to inland. There are very little changes in the wind speed at the distances from 20 to 60 km from the coast to the sea.

Table 4 gives the summary statistics of wind speed characteristics at 100 m for 4 sites with different distance from the seacoast. Site I is located in a coastal area very close to the sea. Another three (II–IV) sites are located in offshore sites and present the locations for offshore wind farms in the near future.

**TABLE 4:** Wind statistics at 100 m for 4 sites located in different distance from the sea coast (1981–2010).

| Site | Coordinates | | | Wind speed, m/s | CV, % | Power density, W/m² |
|------|---------|------|------------|-----------------------|--------------------|---------------------|
|      | Long    | Lat  | Dist, km   | Annual Winter/summer  | Annual Monthly     |                     |
| I (198) | 56.4834 | 21.0166 | −1 | 7.9 | 3 | 460 |
|         |         |         |    | 8.6/7.0 | 9 (VI) | |
|         |         |         |    |         | 16 (IV, IX) | |
| II (367) | 56.9325 | 20.9391 | 7 | 8.8 | 3 | 647 |
|          |         |         |   | 10.0/7.6 | 10 (XII) | |
|          |         |         |   |          | 17 (IV, IX) | |
| III (160) | 56.3937 | 20.6858 | 15 | 8.9 | 3 | 671 |
|           |         |         |    | 10.2/7.6 | 10 (I, VI) | |
|           |         |         |    |          | 18 (IV) | |
| IV (362) | 56.9322 | 20.5285 | 26 | 9.1 | 3 | 711 |
|          |         |         |    | 10.3/7.9 | 10 (I, VI) | |
|          |         |         |    |          | 18 (IV) | |

*CV: coefficient of variation is calculated as standard deviation/mean * 100.*

Variations in the annual wind speed and related wind energy density are also of great importance for the wind farm development. According to 30-year CLM datasets, the range of annual wind speed indicates that the annual wind speed for any given year within a 30-year period (1981–2010) may be over 3% higher or lower than the mean wind speed during the period. Higher degree of interannual variability was found for monthly average wind speed: up to 20% in spring (April) and autumn (September) and up to 10% in summer.

For calm to weak winds, the wind turbines are at halt. The wind speed at which the turbine first starts to rotate and to generate a power is typically between 3 and 4 m/s. To understand consistency of electrical output, we examined the persistence of wind speeds exceeding 4 m at different locations and different heights. Annual mean percent active (speed > 4 m/s) daily values at a 100 m height range from 92 to 95% for all of the offshore sites, while the terrestrial sites at 100 m height are active 88 to 90% of the time. Table 5 illustrates the annual mean percentage of days with active wind speed for four sites at different heights. The results show that for the offshore territories there are no significant differences in the occurrence of active wind speed at the heights 80–100 m. The occurrence of active wind speed increases noticeably between 10 m and 80 m height particularly for the terrestrial site.

**TABLE 5:** Annual mean percentage of days with active wind speed (daily mean wind speed > 4 m/s).

| Site | 10 m | 80 m | 90 m | 100 m |
|------|------|------|------|-------|
| I    | 73   | 91   | 92   | 92    |
| II   | 85   | 93   | 94   | 95    |
| III  | 86   | 94   | 95   | 96    |
| IV   | 89   | 94   | 95   | 96    |

*The location characteristics of the sites are given in Table 4.*

In general, the results of wind speed mapping and wind power potential assessment indicated that offshore wind resources in the territory are promising to expand national electricity generation.

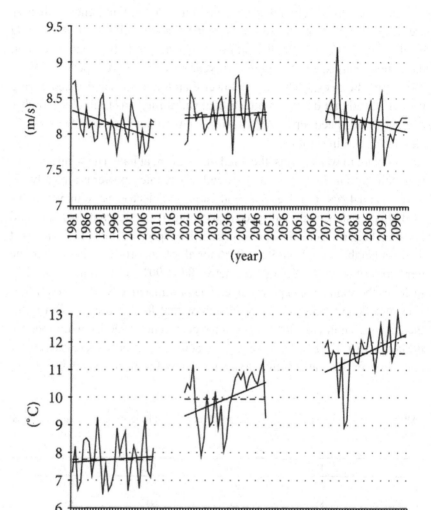

**FIGURE 5:** Time series of annual wind speed (a) and air temperature (b) for a point in the Baltic Sea (56.3937N, 20.6858E). Solid black lines show the tendencies for the respective time series. Dashed line reflects the 30-year average for the each period respectively.

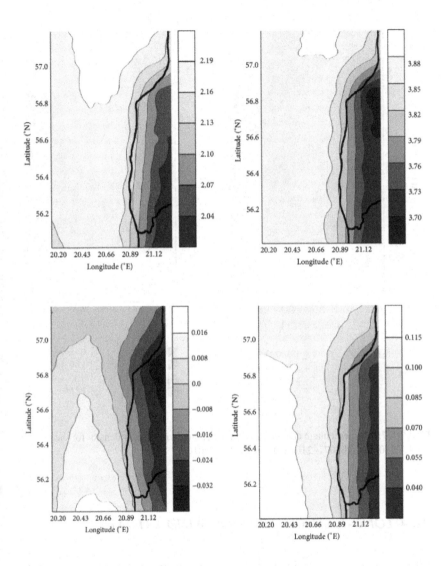

**FIGURE 6:** (Images start in the top left and move clockwise.) (a) Changes in the mean air temperature between 1981–2010 and 2021–2050. (b) Changes in the mean air temperature between 1981–2010 and 2071–2100. (c) Changes in the annual average wind speed between 1981–2010 and 2021–2050. (d) Changes in the annual average wind speed between 1981–2010 and 2071–2100.

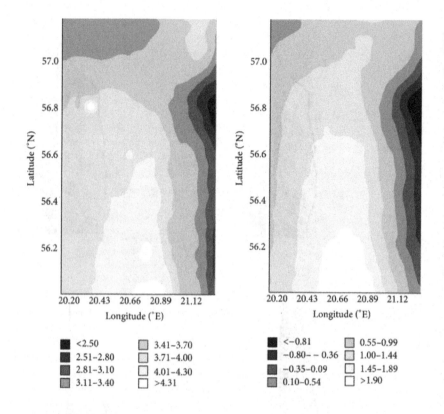

**FIGURE 7:** The percentage change in averaged annual wind power density for the periods 2021–2050 (a) and 2071–2100 (b) versus 1981–2010.

## 2.3.4 LONG-TERM CHANGES OF WIND CLIMATE

Like many other renewable technologies wind energy is also susceptible to climate change. The principal and most direct mechanism by which global climate change may impact the wind energy industry is by changing the geographical distribution and/or the inter- and intra-annual variability of the wind resources [18]. The analysis of the changes in long-term historical wind speed data was beyond the scope of this study. However it should be mentioned that the trend analysis of the average wind speed and maximum wind gusts discloses the decreasing trends in coastal observation stations Liepaja 1 and Pavilosta. In general the results of the changes in the wind climate from two costal weather stations were compared with the data of other 23 meteorological stations in the territory of Latvia. Most of these stations also showed a decreasing tendency of the wind speed and at the same time no significant changes in the wind direction. The analysis of the changes in wind showed that it is very difficult to create homogeneous wind data series even for a comparatively short time periods. Changes in local environment at the vicinity of the meteorological stations (growing of trees and buildings) present the main reason of inhomogeneities in wind data series in the territory of Latvia. However, a similar pattern of surface winds stilling in general has been found in the North Hemisphere [17]. This is consistent with the tendencies found in Europe, Australia, and USA. The studies that have analysed the wind speed data from terrestrial anemometers have generally found declines over the last 30–50 years [18–20], the cause of which is currently uncertain. The long-term changes of wind speed and direction in the territory of Latvia should be evaluated in the future taking into account the possible inhomogeneities.

The same decreasing tendency of wind speed over the sea for the period 1981–2010 has been shown by the CLM simulations. The historical trends of the wind speed were calculated for each grid cell in the study region. The results for one sample grid cell are presented in Figure 5.

Air density affects the energy density in the wind and hence the power output of wind turbines and is inversely proportional to air temperature, and thus increasing air temperature will lead to slight declines in air density and power production in the future. The effect is modest, but not negligible. It is found that at mean sea-level pressure an increase in air temperature by 5°C leads to a decrease in air density by 1-2% with a commensurate decline in energy density [18]. Downscaled results for the middle of the 21st century (2021–2050) exhibit a very significant increase in the air temperature from 2°C to 2.2°C in the covered territory (Figure 6(a)). However, very little differences (higher values) in the annual average wind speed were found in 2021–2050 relative to 1981–2010 (Figure 6(c)). Generally, the downscaled results for the end of the twenty first century (2071–2100) for a given regional climate model indicate higher magnitude changes of air temperature than those from the middle of the 21st century. The annual mean air temperature in the region will be from 3.7°C to 4.0°C higher than in 1981–2010 (Figure 6(b)). In 2071–2100 all grid points exhibit lower values of the mean wind speed relative to 2021–2050 and very little differences in the annual average wind speed relative to 1981–2010 (Figure 6(d)).

Figure 7 shows the projected percentage changes of 100 m annual mean wind speed for the periods 2021–2050 and 2071–2100 versus control period 1981–2010.

Small changes (2–4%) were observed in the energy content of the wind for the middle of the 21st century. For the end of the 21st century the projections show an increase of the wind power density by 1-2% for offshore territories and neglected decrease (less than 1%) for coastal areas.

It should be noted that the projections of the mean wind speed (and wind power) outlined in this paper should be viewed with caution since the climate change signal is of similar magnitude to the variability of the evaluation and control simulations. Also, the climate change projections are subject to a degree of uncertainty that limits their utility. Accordingly, future work should be focused on employing more models.

## 2.4 CONCLUSIONS

The regional climate model (CLM) and High Resolution Limited Area Model (Hirlam) simulations have been used to evaluate the wind power resources in the target territory. The results show that the model data are complementary to observations obtained from the meteorological stations and satellites and can therefore be used to usefully map long-term wind resources in large extensional offshore areas. The created maps provide a better estimate of the offshore wind resources than the previously available evaluation using the in situ observation data from the coastal stations. For offshore territories the wind power resources are predicted to be more than 500 W/m². The results indicated that the offshore wind resources are promising for expanding national electricity generation and reducing the country's air emissions. A significant climate change signal—an increase in the air temperature—was found for the periods 2021–2051 and 2071–2100 relative to the climate reference period 1981–2010. Such shifts in the thermal regimes will likely be related to a decrease in the icing conditions and will result in less extreme cold conditions that will be the major advantage for installation, operation, and maintenance of wind turbines in the future. The recent studies [19, 20] emphasize the effects global warming may have on the harvesting of wind for energy production. According to the authors the wind energy resource may shrink in the future as climate warms in the large territories of Europe. The studies show the slight decrease of wind power output in most regions analyzed. In our study the different assessment method in smaller territory was used compared to these studies. However our data shows the same results: no significant changes in the average wind speed and interannual variability of annual and monthly wind speeds were found for the middle and end of the 21st century relative to 1981–2010. This means that wind energy resources will not change significantly during the 21st century. This work suggests that wind energy will continue to be a stable resource for electricity generation in the region over the 21st century. The future work should be focused on using more climate models as well as emission scenarios for assessing the possible future wind climate and wind power potential changes.

## REFERENCES

1.  Europe's Onshore and Offshore Wind Energy Potential, "An assessment of environmental and economic constraints," Tech. Rep. no. 6, EEA, 2009.
2.  L. Landberg, L. Myllerup, O. Rathmann et al., "Wind resource estimation—an overview," Wind Energy, vol. 6, no. 3, pp. 261–271, 2003.
3.  A. M. Sempreviva, R. J. Barthelmie, and S. C. Pryor, "Review of methodologies for offshore wind resource assessment in European seas," Surveys in Geophysics, vol. 29, no. 6, pp. 471–497, 2008.
4.  C. Hasager, A. Pena, M. Christiansen et al., "Remote sensing observation used in offshore wind energy," IEEE Journal of Selected Topics in Applied Earth Observations and Remote Sensing, vol. 1, pp. 67–79, 2008.
5.  R. J. Barthelmie and S. C. Pryor, "Can satellite sampling of offshore wind speeds realistically represent wind speed distributions?" Journal of Applied Meteorology, vol. 42, pp. 83–94, 2003.
6.  B. R. Furevik, A. M. Sempreviva, L. Cavaleri, J.-M. Lefèvre, and C. Transerici, "Eight years of wind measurements from scatterometer for wind resource mapping in the Mediterranean Sea," Wind Energy, vol. 14, no. 3, pp. 355–372, 2011.

7.   S. C. Pryor, R. J. Barthelmie, and E. Kjellström, "Potential climate change impact on wind energy resources in northern Europe: analyses using a regional climate model," Climate Dynamics, vol. 25, no. 7-8, pp. 815–835, 2005.

8.   P. Nolan, P. Lynch, R. Mcgrath, T. Semmler, and S. Wang, "Simulating climate change and its effects on the wind energy resource of Ireland," Wind Energy, vol. 15, no. 4, pp. 593–608, 2012.

9.   ENSEMBLES, "Ensemble-based prediction of climate change and their impacts," 2006, http://www.ensembles-eu.org/.

10.  ENSEMBLES, "Climate change and its impacts at seasonal, decadal and centennial timescales. Summary of research and results from the ENSEMBLES project," 2009, http://ensembles-eu.metoffice.com/docs/Ensembles_final_report_Nov09.pdf.

11.  C. D. Hewitt and D. J. Griggs, "Ensembles-based predictions of climate changes and their impacts," Eos, vol. 85, no. 52, p. 566, 2004. View at Scopus

12.  D. Cepīte, U. Bethers, A. Timuhins, and J. Seņņikovs, "Penalty function for identification of regions with similar climatic conditions," in Climate Change and Latvia, pp. 8–16, University of Latvia, Riga, Latvia, 2011.

13.  J. Sennikovs and U. Bethers, "Statistical downscaling method of regional climate model results for hydrological modelling," in Proceedings of the 18th World IMACS / MODSIM Congress, Cairns, Australia, 2009, http://www.mssanz.org.au/modsim09/I13/sennikovs.pdf.

14.  N. Nakicenovic and R. Swart, Eds., Special Report on Emissions Scenarios. A Special Report of Working Group III of the Intergovernmental Panel on Climate Change, Cambridge University Press, Cambridge, UK, 2000.

15.  T. Thorsteinn and H. Björnsson, Eds., Climate Change and Energy Systems. Impact, Risks and Adaptation in the Nordic and Baltic Countries, TemaNord, 2011.

16.  "Operational Meteorological and Oceanographycal system for Baltic Sea—FIMAR," University of Latvia, 2009.

17.  K. Fennig, A. Andersson, S. Bakan, C. P. Klepp, and M. Schröder, "Hamburg Ocean Atmosphere Parameters and Fluxes from Satellite Data—HOAPS 3. 2—Monthly Means/6-Hourly Composites," Satellite Application Facility on Climate Monitoring, 2012.

18.  S. C. Pryor and R. J. Barthelmie, "Climate change impacts on wind energy: a review," Renewable and Sustainable Energy Reviews, vol. 14, no. 1, pp. 430–437, 2010.

19.  D. Ren, "Effects of global warming on wind energy availability," Journal of Renewable and Sustainable Energy, vol. 2, no. 5, Article ID 052301, 2010.

20.  I. Barstad, A. Sorteberg, and M. D.-S. Mesquita, "Present and future offshore wind power potential in northern Europe based on downscaled global climate runs with adjusted SST and sea ice cover," Renewable Energy, vol. 44, pp. 398–405, 2012.

# CHAPTER 3

# Power Generation Expansion Planning Including Large Scale Wind Integration: A Case Study of Oman

ARIF S. MALIK AND CORNELIUS KUBA

## 3.1 INTRODUCTION

In the planning of a power system, it is essential to estimate the operating cost and reliability of the system. To make these estimations, it is important to model the system load and generation units in an appropriate way. Power system planning is made up of the electrical load forecast, generation planning, and electrical network planning [1, 2]. The electrical load forecast forms the basis of power system planning and provides information on expected consumption increase, load curve profiles, and load distribution. The result of generation planning and electrical network can also conversely exert an influence on electrical load curve or distribution via marginal cost effort. In the planning process, major decisions in expansion planning of the generation system must consider alternative generating unit sizes, types of capacity, timing of addition, and locations. The main sources of uncertainty in strategic planning are forecasts of electricity demand, fuel prices and availability, availability and performance of new

*Power Generation Expansion Planning Including Large Scale Wind Integration: A Case Study of Oman.*
© *Malik AS and Kuba C.* Journal of Wind Energy *2013 (2013). http://dx.doi.org/10.1155/2013/735693.*
*Licensed under a Creative Commons Attribution 3.0 Unported License, http://creativecommons.org/licenses/by/3.0/.*

technology, governmental policies toward privatization and regulations, and public attitudes [3].

This paper reports a study that was aimed to find a least-cost generation expansion plan for the Main Interconnected System (MIS) of Oman considering the large scale integration of wind energy. The importance of the study is to show how wind turbines can be modeled in generation expansion planning software models that are based on load duration curve technique and that wind energy system can form part of the least-cost plan in Oman while keeping the same planning standard of minimum reserve margin and cost of unserved energy.

The study is limited by its scope as no detailed investigation of wind energy sites and their true potential is estimated. However, ample literature on wind energy potential and its application in Oman is available which suggests that there is a significant potential available in the southern part and the coastal area of the country [4–10]. The study also does not consider the cost related to wind power evacuation from the south to the north of the country where the major load exists. The load and generation data used is taken from Oman Power and Water Procurement (OPWP) company's report [11] and private communications to OPWP personnel. The candidate plants used for expanding the generation system are taken those which are presently in the Omani power system with the exception of wind plants. The generation expansion planning is done using Wien Automatic System Planning (WASP) software [12]. The software has its own modeling limitations; for example, the load model is based on the load duration curve (LDC). LDC model provides information about the percentage of time the load equals or exceeds a certain MW value and gives the same energy information as the actual chronological load curve; however, the chronology of the events is lost which poses modeling limitations to nondispatchable technologies such as wind.

This paper is arranged in six sections. Section 1 is an introduction. Section 2 is a theoretical review of power planning concepts. Section 3 is a review of wind energy potential in Oman. Section 4 discusses how wind plants can be modeled in WASP. Section 5 provides generation and load data of MIS system. Section 6 presents the results of the generation expansion planning. And the last section concludes the paper.

## 3.2 ELECTRIC POWER SYSTEM PLANNING

### 3.2.1 OBJECTIVE OF POWER PLANNING STUDY

The objective of electric planning study is to meet the load forecast with high reliability at a minimum cost. There are three keywords in the previous statement, that is, load forecast, reliability, and cost. A brief discussion on these three keywords is followed. The cost is minimized depending on the financial resources, technical, environmental, and political considerations. Four questions must be answered when the study of capacity planning is done [1].

1. What type of capacities will be added to the system?
2. How much capacities will be added?
3. When these capacities will be added?
4. Where these capacities will be located?

The first three questions can be answered by using any generation expansion planning software. However, for the question where to locate the new facilities a detailed feasibility study has to be carried out considering the load center, availability of fuel, water, manpower, transmission corridors, and so forth. The objective function of the least-cost-planning software is normally the following:

Minimize for all j,

$$B_j = \sum_{i=1}^{n} (\overline{CC_i} - \overline{SV_i} + \overline{FC_i} + \overline{O_i\&M_i} + \overline{UE_i})$$

(1)

where

- CC is the capital cost;
- SV is the salvage value;
- FC is the fuel cost;

- O&M is the operating and maintenance cost;
- UE is the energy not served cost;
- n is number of years in study period;
- And over bar on the above costs represents present worth of all the costs.

The capital costs of only candidate generating plants considered for expansion are added in the objective function. The capital costs of existing plants and those in the construction phase (committed plants) in the system are considered sunk and are thus not considered in the objective function.

### 3.2.2 GENERATION PLANNING STUDY PERIOD

The study period normally spans from twenty to twenty-five years. The study period consists of three subperiods: preplanning period, planning period, and postplanning period [13]. Preplanning period is the first 3-4 years in which the planning was done earlier. This is included in the study period to see the energy production cost and reliability of the system. The planning period is between 4 and 10 years in which the decision has to be taken now about the plants that have to be added in the future to meet the forecast. The post-planning period is added in the study period to avoid the end effects by making the calculations continue for another 10 years or more so that a proper tradeoff between construction costs and operating costs is found. Therefore, the long-term forecasts of 20–25 years of electricity consumption and demand are used in the planning of investment in generating capacity and the development of fuel supplies.

### 3.2.3 RELIABILITY

Reliability is the ability of a system or component to perform its required functions under stated conditions for a specified period of time [14, 15]. In (1) the reliability is also embedded in terms of cost of energy not served. This cost is basically customers' electricity outages cost and is the penalty for unreliability. If more capacity is added in the system, the capital cost would be high but because of more capacity in the system the total unserved energy cost would be low as the system would be more reliable. On the other hand, if the capacity added is less than the actual required, then there would be more customer outages and hence higher unserved energy cost. The basic concept of reliability-cost/reliability-worth evaluation is relatively simple and can be presented by the cost/reliability curves, as shown in Figure 1 [14, 15]. These curves show that the investment cost generally increases with higher reliability, that is, capital investment is augmented in order to improve the reliability. On the other hand, the customers' outages cost decreases as the reliability increases. Furthermore, the total cost of the two curves is the sum of the utility cost and the customers cost, and the minimum exhibits the optimal cost.

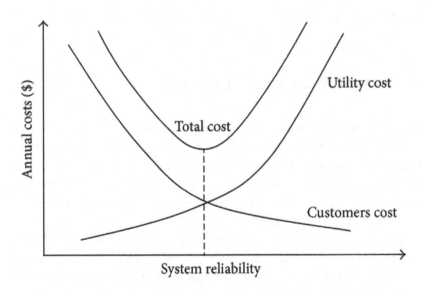

**FIGURE 1:** Relationship between consumers, utility, and total annual cost with respect to reliability.

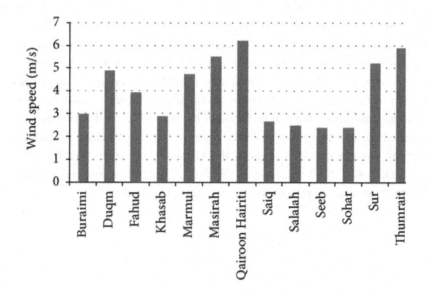

**FIGURE 2:** Annual average wind speeds of some sites in Oman at 10 m height.


## 3.3 WIND ENERGY POTENTIAL IN OMAN

The wind speed in Oman is relatively high compared with other gulf countries. Oman's southern region appears to have the highest wind potential. Figure 2 shows the average wind speed in some of the cities in Oman. The highest wind potential is in Quiroon Hariti and Thumrait and both are in the South of Oman. A brief literature review on wind energy potential in Oman is presented in the following paragraphs.

In [6], it is concluded that with the existing gas price of 1.5 US\$/MMBtu wind energy is not economical for grid application. The wind energy at Quiroon Hariti, (see Figure 2) the highest wind potential in Oman, becomes marginally economical at a gas price of 6 \$/MMBtu. At the opportunity cost of natural gas price of approximately 3 \$/MMBtu and adding a depletion premium of 3% per annum, the cost of wind energy become comparable to open cycle gas-turbine (GT) power plants. The combined cycle (CCGT) power plants remain cheaper, however. The comparison is made by assuming the economic life of assets (GT, CCGT, and 20 MW wind farm) to be 25 years and the real discount rate at 7.55%.

In [7], the electricity generation from wind energy for Duqm, a coastal region, was investigated based on the monthly mean wind speed observations. A technoeconomic evaluation was also presented using V90-1.8 turbine. It was concluded that the power generation cost is higher than the current existing system, due to the highly subsidized price of natural gas. In [8], a single 50 kW wind turbine of TekVal was used to demonstrate the economical utilization of the wind energy at the site. It was concluded that the operating cost of the diesel generation was 1.7–1.8 times the specific cost of wind turbine. It was also concluded that the simple payback period of the turbine was about five years.

In [9], five-year hourly wind data is analyzed from twenty-nine weather stations to identify the potential location for wind energy applications in Oman. Different criteria including theoretical wind power output, vertical profile, turbulence, and peak demand fitness were considered to identify the potential locations. Air density and roughness length, which play an important role in the calculation of the wind power density potential, are derived for each station site. Due to the seasonal power demand, a seasonal approach is also introduced to identify the wind potential on different seasons. Finally, a scoring approach was introduced in order to classify the potential sites based on the different factors mentioned previously. It is concluded that Qairoon Hairiti, Thumrait, Masirah, and Ras Alhad have high wind power potential and that Quiroon Hariti is the most suitable site for wind power generation.

In [10], the article assessed wind power cost per kWh of energy produced using four types of wind machines at 27 locations within Oman. These sites cover all regions in Oman. Hourly values of wind speed recorded between 2000 and 2009, in most cases, were used for all 27 locations. Wind duration curves were developed and utilized to calculate the cost per kWh of energy generated from four chosen wind machines. It was found that the cost of energy is low in the south and middle regions of Oman compared with that in the north region. According to the study, the most promising sites for the economic harnessing of wind power are

Thumrait, Qairoon Hairiti, Masirah, and Sur, with an energy cost of less than 0.117 US$/kWh when 2000 kW, 1500 kW, 850 kW, or 250 kW wind turbines are used.

## 3.4 WIND PLANT MODELING IN WASP

As mentioned in the introduction that the load model in WASP software is based on the load duration curve (LDC) technique. LDC model provides information about the percentage of time the load equals or exceeds a certain MW value and gives the same energy information as the actual chronological load curve. However, the chronology of the events is lost in LDC models which pose modeling limitations for nondispatchable technologies such as wind which are inherently chronological devices that operate according to the availability of wind. There are several ways to model wind turbine/plant in WASP and all have some kind of approximation. Here are some of the ways to handle it.

1. Wind turbine/plant can be modeled as a thermal plant with zero fuel cost and increased forced outage rate to reflect the variability of wind and reduced capacity credit of wind turbine. With zero fuel cost, the economic loading order of wind turbine is first in the merit and can be considered as a base load plant.
2. Wind turbine is modeled as negative load. The expected energy produced by a wind turbine can first be subtracted from the original chronological load curve and then the load duration curve is made. The optimization is then done without considering wind turbines and the cost of wind turbine can then be added later in the optimal case.
3. Wind turbine can be derated according to the capacity credit and modeled as a thermal plant with zero fuel cost and a normal forced outage rate of say about 4% can be assigned.
4. Wind turbine is modeled as a hydroplant with a base load capacity and inflow energy as a constraint. The inflow energy reflects the energy the wind turbine can produce in a given load duration curve time span.

In the present study, a wind turbine is modeled as a thermal plant with high forced outage rate to investigate the economic feasibility of wind plants. This technique is closer to reality as the higher forced outage rate force the WASP model to select more plants to meet the reliability requirement. In real systems also because of poor capacity credit attributed to wind energy a lot of backup supply has to be provided. The candidate plant of 2 MW taken from [10] is used to demonstrate the economic feasibility at one of the best sites of Thumrait. Table 1 gives the technical data of wind turbine.

**TABLE 1:** Technical data of wind turbine.

| Wind mach. | Rated power (kW) | Cut-in (m/s) | Cut-out (m/s) | Rated speed (m/s) | Hub height (m) | Rotor Dia (m) | Expected life (yrs) |
|---|---|---|---|---|---|---|---|
| V80 | 2000 | 4 | 25 | 16 | 67 | 80 | 20 |

**TABLE 2:** Data of wind turbine for WASP.

| Type | Max. capacity (MW) | Forced outage rate | Sch. main-tenance days | Fixed O&M cost $/kW-month | Capital cost $/kW | Life (yrs) | Construc-tion time (yr) |
|---|---|---|---|---|---|---|---|
| Wind | 20 | 75.4% | 10 | 1.46 | 1500 | 20 | 1 |

The annual energy produced by this wind turbine at the site is 4318 MWh from which its forced outage rate is calculated from its capacity factor using the following two formulas:

$$\text{Capacity Factor} = \frac{\text{Annual Energy Produced (MWh)}}{\text{Rated Capacity (MW)} \times 8760 \text{ hrs}}$$

$$\text{Forced Outage Rate} = 1 - \text{Capacity Factor} \tag{2}$$

It is assumed that 10 wind turbines of 2 MW size form a wind park of 20 MW. This 20 MW is used as a candidate plant instead of 2 MW wind turbine. The other needed data for WASP is shown in Table 2.

## 3.5 MIS GENERATION AND LOAD DATA

### 3.5.1 GENERATION DATA

The generation expansion planning is carried out for the MIS from 2012 to 2034. The main Interconnected System consists of ten existing plants Ghubrah, Rusail, Wadi Al-Jizzi, Manah, Barka-I, Barka-II, Barka-III, Sohar-I, Sohar-II, and Alkamil. All the plants are either combustion turbines or combined cycle plants. The fixed or existing system consists of about 4770 MW capacity [11]. The committed plants and the retirements are also taken into account. Table 3 provides technical and economic data for the fixed system and committed plants. All power plants are powered by natural gas, which comes from domestic production with a fuel cost of 1189 ¢/$10^6$ kcal equivalent to \$3/MMBtu. The last two plants with zero number of sets show the committed plants. Table 4 provides the technical and economic data of units taken as candidate plants for expansion.

### 3.5.2 LOAD DATA

As mentioned earlier, the load model in WASP is of load duration curve. The annual chronological hourly load curve of year 2009 is used to make load duration curves (LDC) for winter and summer seasons. Figure 3 shows a summer load duration curve with inverted axes. The figure shows that load of 1100 MW or more needs to be served all times and more than 3000 MW needs to be served for around 8% of time. These summer and winter LDCs are then normalized and the shapes of LDCs are assumed the same for the whole study period. The peak load for the 2012 is 4189 MW and for year 2034 (the end of study period) is 12,617 MW with an average load growth of about 5%. Figure 4 shows the peak load from 2012 till 2034.

**TABLE 3:** Technical and cost data of thermal plants in the year 2012.

| No. | Plant name | No. of sets | Min. load MW | Ca-pacity MW | Heat rates kcal/kWh | | Fast spin Res % | FOR % | Sched. Main-tenance days | O&M (FIX) $/kW-month | O&M (VAR) $/ MWh |
|---|---|---|---|---|---|---|---|---|---|---|---|
| | | | | | Base load | Incr. load | | | | | |
| 1 | GBG5 | 8 | 6 | 16 | 4728 | 2434 | 9 | 10 | 27 | 3.78 | 0.18 |
| 2 | GBG6 | 2 | 11 | 26 | 4176 | 2116 | 9 | 10 | 31 | 5.21 | 0.18 |
| 3 | GBG9 | 2 | 91 | 91 | 2705 | 2705 | 0 | 5 | 31 | 7.17 | 0.18 |
| 4 | GBS1 | 1 | 37 | 37 | 1811 | 1811 | 0 | 50 | 34 | 5.68 | 0.37 |
| 5 | GBS2 | 2 | 30 | 30 | 1058 | 1058 | 0 | 20 | 25 | 8.91 | 0.37 |
| 6 | RUSL | 8 | 45 | 86 | 3713 | 2224 | 9 | 5 | 27 | 3.57 | 2.26 |
| 7 | WJZ1 | 8 | 12 | 27 | 4204 | 2205 | 9 | 5 | 31 | 6.50 | 0.18 |
| 8 | WJZ2 | 5 | 6 | 17 | 4701 | 2430 | 9 | 5 | 30 | 6.50 | 0.18 |
| 9 | MNH1 | 3 | 12 | 29 | 4301 | 2253 | 9 | 2 | 20 | 6.52 | 3.24 |
| 10 | MNH2 | 2 | 42 | 94 | 3455 | 2215 | 9 | 2 | 30 | 6.52 | 2.13 |
| 11 | ALK | 3 | 44 | 99 | 3515 | 2326 | 9 | 2 | 37 | 6.57 | 3.32 |
| 12 | BRK1 | 1 | 108 | 435 | 3257 | 1756 | 9 | 3 | 20 | 9.08 | 1.44 |
| 13 | BRK2 | 1 | 126 | 710 | 2114 | 1742 | 9 | 4 | 20 | 7.07 | 1.16 |
| 14 | BRK3 | 2 | 40 | 247 | 1797 | 1455 | 9 | 3 | 20 | 20.05 | 1.04 |
| 15 | SHR1 | 1 | 121 | 590 | 3467 | 1821 | 9 | 3 | 20 | 5.84 | 1.28 |
| 16 | SHR2 | 2 | 40 | 247 | 1797 | 1455 | 9 | 3 | 20 | 20.05 | 1.04 |
| 17 | SURa | 0 | 400 | 799 | 1771 | 1316 | 9 | 3 | 20 | 13.01 | 0.66 |
| 18 | SURb | 0 | 201 | 401 | 1711 | 1316 | 9 | 3.5 | 20 | 13.01 | 0.66 |

## 3.5.3 ECONOMIC DATA

The discount rate used for the study is 7.5% and the cost of unserved energy as 1.6 $/kWh. The discount rate of 7.5% has been used in earlier official studies; see for example [6]. The cost of unserved energy was worked out in [15] and is consistent with the optimal reserve margins found using the cost of unserved energy and the average reserve margins existing in Oman.

**TABLE 4:** Technical and cost data of candidate plants.

| No. | Plant name | Min. load MW | Capacity MW | Heat rates kcal/kWh | | Fast spin Res % | FOR % | Sched. maintenance days | O&M (FIX) $/kW-month | O&M (VAR) $/MWh | Capital cost $/kWh | Plant life (yrs) | Construction time |
|-----|------------|--------------|-------------|---------------------|--|-----------------|-------|-------------------------|----------------------|-----------------|--------------------|------------------|-------------------|
| | | | | Base load | Incr load | | | | | | | | |
| 1 | GBG9 | 91 | 91 | 2705 | 2705 | 0 | 5 | 31 | 7.17 | 0.18 | 795 | 25 | 3 |
| 2 | BRK1 | 108 | 435 | 3257 | 1756 | 9 | 3 | 20 | 9.08 | 1.44 | 977 | 25 | 3 |
| 3 | FCOL | 83 | 250 | 2800 | 2300 | 10 | 8 | 35 | 2.92 | 5.00 | 2000 | 25 | 5 |

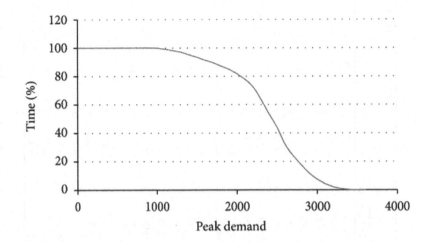

**FIGURE 3:** Load duration curve with inverted axis.

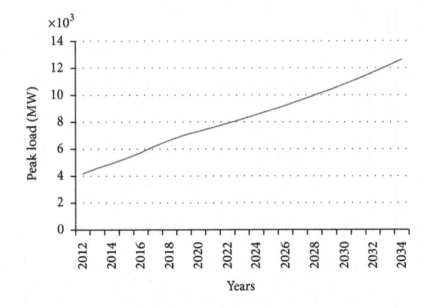

**FIGURE 4:** Annual peak demand from 2012 to 2034.

**TABLE 5:** Results of base case: costs, reliability, type, and number of units selected.

| Year | Present worth cost of the year in thousands dollars | | | | | Obj. Fn (Cumulative) | LOLP % | GBG9 | BRK1 | FCOL |
|---|---|---|---|---|---|---|---|---|---|---|
| | Const. cost | Salvage value | Operating cost | Energy not served cost | Total | | | | | |
| 2034 | 81646 | 73253 | 631261 | 0 | 639654 | 21847338 | 0.028 | 74 | 15 | 0 |
| 2033 | 104834 | 84241 | 651639 | 0 | 672232 | 21207684 | 0.024 | 69 | 15 | 0 |
| 2032 | 109827 | 78893 | 670504 | 0 | 701438 | 20535452 | 0.031 | 63 | 15 | 0 |
| 2031 | 117515 | 75307 | 689700 | 0 | 731908 | 19834014 | 0.028 | 63 | 14 | 0 |
| 2030 | 826712 | 471547 | 709039 | 0 | 1064204 | 19102106 | 0.027 | 63 | 13 | 0 |
| 2029 | 337793 | 171065 | 699590 | 126 | 866444 | 18037902 | 0.047 | 42 | 10 | 0 |
| 2028 | 143960 | 64548 | 734243 | 123 | 813778 | 17171458 | 0.046 | 39 | 8 | 0 |
| 2027 | 104885 | 41510 | 752857 | 235 | 816467 | 16357680 | 0.055 | 39 | 7 | 0 |
| 2026 | 112227 | 39068 | 777024 | 262 | 850445 | 15541213 | 0.055 | 35 | 7 | 0 |
| 2025 | 412757 | 125894 | 802078 | 327 | 1089268 | 14690768 | 0.058 | 31 | 7 | 0 |
| 2024 | 188703 | 50204 | 812510 | 372 | 951381 | 13601500 | 0.057 | 29 | 5 | 0 |
| 2023 | 708848 | 163664 | 830018 | 796 | 1375999 | 12650119 | 0.085 | 29 | 4 | 0 |
| 2022 | 252822 | 50358 | 823948 | 867 | 1027279 | 11274120 | 0.086 | 26 | 1 | 0 |
| 2021 | 157404 | 26859 | 846190 | 1015 | 977749 | 10246841 | 0.092 | 25 | 0 | 0 |
| 2020 | 252633 | 36626 | 867924 | 1639 | 1085570 | 9269092 | 0.124 | 21 | 0 | 0 |
| 2019 | 450528 | 54940 | 893340 | 1709 | 1290638 | 8183523 | 0.121 | 15 | 0 | 0 |
| 2018 | 241033 | 24418 | 910721 | 2313 | 1129649 | 6892885 | 0.144 | 5 | 0 | 0 |
| 2017 | 0 | | 933210 | 535 | 933745 | 5763236 | 0.047 | 0 | 0 | 0 |
| 2016 | 0 | 0 | 961424 | 0 | 961424 | 4829492 | 0.006 | 0 | 0 | 0 |
| 2015 | 0 | 0 | 996447 | 0 | 996447 | 3868068 | 0 | 0 | 0 | 0 |
| 2014 | 0 | 0 | 997687 | 0 | 997687 | 2871621 | 0.001 | 0 | 0 | 0 |
| 2013 | 0 | 0 | 965412 | 965 | 966377 | 1873934 | 0.071 | 0 | 0 | 0 |
| 2012 | 0 | 0 | 906425 | 1132 | 907557 | 907557 | 0.076 | 0 | 0 | 0 |

**TABLE 6:** Results of base case: costs, reliability, type, and number of units selected.

| Year | Present worth cost of the year in thousands dollars | | | | | | LOLP% | GBG9 | BRK1 | Wind |
|---|---|---|---|---|---|---|---|---|---|---|
| | Const. cost value | Salvage value | Operating cost | Energy not served cost | Total | Obj. Fn (cumulative) | | | | |
| 2034 | 112256 | 100716 | 604143 | 1883 | 617566 | 21721956 | 0.284 | 63 | 15 | 51 |
| 2033 | 113361 | 90334 | 619355 | 2361 | 644744 | 21104390 | 0.328 | 62 | 14 | 51 |
| 2032 | 125332 | 89652 | 640912 | 1968 | 678560 | 20459646 | 0.276 | 58 | 14 | 45 |
| 2031 | 125810 | 80370 | 658514 | 2096 | 706050 | 19781086 | 0.282 | 58 | 13 | 43 |
| 2030 | 846035 | 480353 | 675986 | 2353 | 1044020 | 19075036 | 0.299 | 58 | 12 | 42 |
| 2029 | 333885 | 168326 | 669359 | 2367 | 837284 | 18031016 | 0.26 | 39 | 9 | 35 |
| 2028 | 143960 | 64549 | 706327 | 1849 | 787588 | 17193732 | 0.207 | 37 | 7 | 33 |
| 2027 | 154038 | 60963 | 723468 | 2484 | 819027 | 16406144 | 0.25 | 37 | 6 | 33 |
| 2026 | 190940 | 58496 | 740582 | 3493 | 876519 | 15587117 | 0.317 | 37 | 5 | 33 |
| 2025 | 411400 | 124215 | 775462 | 1787 | 1064434 | 14710598 | 0.177 | 36 | 5 | 19 |
| 2024 | 128488 | 34184 | 787557 | 2170 | 884031 | 13646164 | 0.187 | 29 | 4 | 17 |
| 2023 | 739292 | 158541 | 810692 | 2817 | 1394260 | 12762133 | 0.219 | 25 | 4 | 17 |
| 2022 | 216046 | 43033 | 829179 | 1085 | 1003277 | 11367873 | 0.098 | 16 | 3 | 1 |
| 2021 | 231169 | 39447 | 853933 | 792 | 1046448 | 10364596 | 0.076 | 16 | 2 | 1 |
| 2020 | 306917 | 43548 | 869463 | 1793 | 1134626 | 9318148 | 0.131 | 16 | 1 | 1 |
| 2019 | 450528 | 54940 | 893340 | 1709 | 1290638 | 8183523 | 0.121 | 15 | 0 | 0 |
| 2018 | 241033 | 24418 | 910721 | 2313 | 1129649 | 6892885 | 0.144 | 5 | 0 | 0 |
| 2017 | 0 | 0 | 933210 | 535 | 933745 | 5763236 | 0.047 | 0 | 0 | 0 |

## 3.6 RESULTS AND DISCUSSIONS

Table 5 provides the results of a base case without the option of wind turbines. It may be noted that FCOL is not selected at all. The results show that by the year 2034 seventy-four units of GBG9 and fifteen units of BRK1 are selected with total capacity addition of 13,259 MW in the system. The existing generating units of Table 3 will all retire during the study period. The loss-of-load probability (LOLP) of 0.028% in year 2034 corresponds to loss-of-load expectation of about 0.1 day per year or 1 day in 10 years. The total objective function cost is about 21.85 billion dollars.

The result of taking 20 MW wind park as a candidate unit is shown in Table 6. It may be noted that the results are shown from 2017 onward because the results of earlier years are same as the base case. FCOL option was not taken as it was not selected in the base case. The result shows that by the year 2034 sixty-three units of GBG9, fifteen units of BRK1, and fifty one units of Wind (1020 MW) are selected with total capacity addition of 13,278 MW in the system. The total objective function cost is about 21.72 billion dollars which is about 130 million dollars less comparing to the base case. The LOLP in year 2034 is 0.284% about 1 day per year. The reliability of the system is not as good as in the base case but meets the same reserve margins limits of minimum 5% used as a constraint. This also shows that to meet the same reliability criteria of LOLP more capacity additions are required and the reserve margin criteria are not as good measure when intermittent technology are considered in the system. It may also be noted that environmental costs are not considered in the analysis.

## 3.7 CONCLUSIONS

This paper has presented the results of a least-cost generation expansion plan using wind turbines as a candidate plant. A set of 10 wind turbines of 2 MW capacity is considered as a single unit of 20 MW capacity and modeled as a thermal unit in WASP with high forced outage rate according to the expected capacity factor at the selected site. The result shows that with minimum 5% reserve margin reliability criteria and the cost of unserved energy as $1.6/kWh the wind turbine indeed forms part of the generating system expansion economically. However, the result of reliability criteria of LOLP has shown significant difference in the two cases. Although LOLP criteria are superior to reserve margin criteria but in the presence of cost of unserved energy as balancing factor in the objective function LOLP should be automatically taken care of. On the other hand, calculating cost of unserved energy is not an easy task and needs a lot of assumptions and a reasonable sample size of survey in residential, commercial, industrial, and other sectors. As a future work it would be worthwhile to compare the high forced outage rate model of wind turbine with the negative load model. It would also be worthwhile to find the limit of maximum wind potential that can be exploited in the southern part of the country so that it can be added as a constraint to limit the number of wind turbine units selected for strategic plan.

## REFERENCES

1. W. Buehring, C. Huber, and J. Marques de Souza, Expansion Planning for Electrical Generating Systems—A Guidebook, Technical Report Series No. 241, IAEA, Vienna, Austria, 1984.
2. J. Wang and J. McDonald, Modern Power System Planning, McGraw-Hill, London, UK, 1994.
3. J. Stoll, Least-Cost Electric Utility Planning, John Wiley & Sons, New York, NY, USA, 1989.
4. A. S. S. Dorvlo and D. B. Ampratwum, "Summary climatic data for solar technology development in Oman," Renewable Energy, vol. 14, no. 1–4, pp. 255–262, 1998.
5. M. Y. Sulaiman, A. M. Akaak, M. A. Wahab, A. Zakaria, Z. A. Sulaiman, and J. Suradi, "Wind characteristics of Oman," Energy, vol. 27, no. 1, pp. 35–46, 2002.
6. Authority for Electricity Regulation, Oman, "Study on renewable energy eesources, Oman," Final Report, 2008, http://www.aer-oman.org/images/renewables%20 study%20may%202008.pdf.
7. M. H. Albadi, E. F. El-Saadany, and H. A. Albadi, "Wind to power a new city in Oman," Energy, vol. 34, no. 10, pp. 1579–1586, 2009.
8. A. Malik and A. H. Al-Badi, "Economics of wind turbine as an energy fuel saver—a case study for remote application in oman," Energy, vol. 34, no. 10, pp. 1573–1578, 2009.
9. S. AL-Yahyai, Y. Charabi, A. Gastli, and S. Al-Alawi, "Assessment of wind energy potential locations in Oman using data from existing weather stations," Renewable and Sustainable Energy Reviews, vol. 14, no. 5, pp. 1428–1436, 2010.

10. A. H. Al-Badi, "Wind power potential in Oman," International Journal of Sustainable Energy, vol. 30, no. 2, pp. 110–118, 2011.

11. "OPWP's 7-year statement (2102-2018)," Oman Water and Procurement Company, 2012, http://www.omanpwp.com/PDF/Final%207YS%202012-2018.pdf.

12. Wein Automatic System Planning Package, WASP-IV, IAEA, 2003.

13. M. Malone, "Generation planning in the 1980's," in Proceedings of the 11th National Convention of the Institute of Electrical Engineers of the Philippines, Manila, Philippines, November 1986.

14. R. Billinton and R. Allan, Reliability Evaluation of Power System, Pitman, 1996.

15. F. Al-Farsi, M. Al-Shihi, and Y. Al-Shokaili, "Optimal generation reserve margin for Muscat interconnected system," Final Year Project Report, SQU, Electrical Engineering Department, Muscat, Oman, 2000.

# CHAPTER 4

# Developing a GIS-Based Visual-Acoustic 3D Simulation for Wind Farm Assessment

MADELEINE MANYOKY, ULRIKE WISSEN HAYEK,
KURT HEUTSCHI, RETO PIEREN, AND ADRIENNE GRKT-REGAMEY

## 4.1 INTRODUCTION

Planning of new renewable energy installations in the landscape is a complicated matter in Switzerland and all over Europe. Although the public generally supports the renewable energy deployment, the implementation of new installations often fails when it comes to choosing appropriate locations, especially regarding wind farm locations on the local level [1,2]. Cowell [3] points out the "split between the technical and the social" as a key problem: the technical potential is taken as basis for national wind power targets in a top-down approach (e.g., [4]), lacking the public's judgment about the acceptability in particular places. According to recent studies, however, social acceptance is a key issue for successful wind energy market development [5,6]. Furthermore, stakeholders state that there are no suitable instruments to support social acceptance [7].

The impact of the new infrastructures on a specific type of landscape characterized by aesthetical quality and a sense of place is one of the most

Developing a GIS-Based Visual-Acoustic 3D Simulation for Wind Farm Assessment. © Manyoky M, Hayek UW, Heutschi K, Pieren R, and Grêt-Regamey A. ISPRS International Journal of Geo-Information 3,1 (2014), doi:10.3390/ijgi3010029. Licensed under a Creative Commons Attribution 3.0 Unported License, http://creativecommons.org/licenses/by/3.0/.

significant factors explaining support or rejection of wind farms [1,2]. Although the visual sense is the dominant human sensory component for landscape perception, it only provides partial information about our environment [8]. Therefore, a multi-sensory approach for landscape assessment is needed. With regard to wind parks people perceive the noise generated by rotating turbine blades as one of the most prominent annoyance factors [9,10], which is linked to the visual attitude of the wind turbines in the landscape [11]. Hence, there is a strong need for integrating this factor of landscape quality into the site planning of wind farms in order to allow the identification of socially accepted locations for wind power technologies.

### 4.1.1 TECHNOLOGICAL ADVANCES IN DIGITAL LANDSCAPE VISUALIZATIONS

In the last few decades, the technological advances in digital landscape visualization tools and techniques allow landscape and urban planners to use digital 3D visualizations as a common feature for landscape design, planning and management [8]. Highly detailed 3D vegetation for landscape visualizations (e.g., Laubwerk plants [12], XfrogPlants [13] or the SpeedTree Toolkit [14]), integrated augmented reality and Geographic Information System (GIS) representations [15], multi-player capabilities, and communication possibilities [16,17,18] are just a few examples. These advances hold great potential for landscape planners in a diversity of applications such as participatory purposes, communication, scenario evaluation, and the decision-making process [19]. GIS-based 3D visualizations have proved to facilitate the communication between various stakeholders, professionals, and the public in the context of participatory wind power development [20,21]. In the landscape planning context, currently emerging game engines offer interesting modeling tools providing state-of-the-art visualizations [22]. Game engines are software programs including different modules for 2D and 3D representations, and generic physics calculations [23] to design computer games. These engines allow for a high level of interactivity and support multiplayer capabilities with online communications [19]. Some of them are available for low prices in the case of non-commercial use and are designed to run on low budget computers as well [22,24]. One major advantage of game engines is the ability to create real-time visualizations, which do not require a time consuming rendering step, where each camera position and physical parameter has to be set before the visualization can be presented. Instead, real-time visualizations allow the user to move freely through the environment and to dynamically alter physical parameters in the virtual landscape, such as daytime settings or the weather conditions. The ability of game engines to create virtual landscapes with controllable and changeable real world parameters can lead to a suitable planning tool to support communication in decision-making processes. Bishop [19] states that realistic gaming tools configured as a collaborative virtual environment can help people to explore complex issues, run scenario models, and develop acceptable plans. Furthermore, the user's understanding of the real world can be supported by the interactivity and dynamics of virtual environments [25]. If these gaming tools include crucial physical forces and processes of the landscape and also support multi-user capabilities, it would be a logical extension to present such game environments to communities in order to further the design of sustainable future landscapes [19].

## 4.1.2 GAME ENGINES FOR LANDSCAPE VISUALIZATIONS

Using a game engine for landscape visualization and planning, Stock et al. [18] showed how Torque's Game Engine [26] can be employed to establish a web-based map server to create a collaborative environment. Jacobson and Lewis [27] presented an immersive cave-like virtual reality projection of landscapes using the Unreal Engine [28]. Nakevska et al. [29] illustrate examples of a cave-environment using Crytek's CryENGINE [30] interfacing the game engine with sensors and input devices for interacting with the virtual environment. Friese et al. [22] analyzed the visualization and interaction capabilities of different game engines, and decided to use the CryENGINE for modeling landscape visualizations because large outdoor terrains can be generated and edited interactively. The CryENGINE was used by Germanchis et al. [31,32] as well, because the CryENGINE was the most stable, easiest to learn and most powerful engine to generate a virtual environment.

A major advantage of the CryENGINE 3 is that it incorporates a physics engine, which can be applied to almost all objects within the virtual world and allows realistic interaction of objects with physical forces such as wind, gravity, friction, and collisions [30]. Therefore, the engine does not rely on precomputed effects but is capable of displaying physically-based phenomena such as dynamic daytime simulation in real-time, changes in vegetation and cloud movement due to altered wind speed [33].

### 4.1.3 GAME ENGINES AND GEODATA

Germanchis et al. [31,32] uses a workflow to integrate geodata into the game engine transforming the data into an appropriate form for the game environment to understand. The workflow consists of different standalone software programs, e.g., ArcGIS or Photoshop. Herrlich [34] developed a tool to convert real GIS-data into a suitable source readable for the CryENGINE. In the conversion process, Gauss-Krueger coordinates are mapped to game level coordinates. Later, Herrlich et al. [35] show that a game console can be used as a device for geodata visualization and GIS applications, and suggest a way of integrating the standards CityGML and COLLADA for high-quality visualization. However, Friese et al. [22] state that one of the major difficulties encountered during their project using geodata with a game engine were the data conversion processes, as the resolution of the geodata is usually very different compared to the internal engine height map resolution. Furthermore, the coordinate systems of the geodata compared to the internal game engine system vary as well. To get acquainted to the import of terrain, 3D objects and other external data from GIS software, time-intensive trials are often necessary [24].

### 4.1.4 REPRODUCTION OF WIND TURBINE NOISE

With regard to acoustics, the wind turbine noise has to be reproduced for planning future wind farms. Therefore, the audio signals that are suggestive of realistic noise in a given situation have to be synthesized. Sound emission from wind turbines is composed of a mechanical and an aerodynamic component. The mechanical noise is produced by the gearbox and other moving parts of the turbine. The aerodynamic noise on the other hand is generated by air passing the rotor blades. Investigations with microphone arrays have shown that the trailing edge delivers the main contribution to the overall noise [36]. This aerodynamic noise is very broadband with a drop-off towards higher frequencies and can be predicted by numerical calculation methods [37,38,39].

Overall, sophisticated software tools are provided for either landscape visualization or auralization. However, spatially explicit noise emissions of wind turbines integrated in 3D landscape visualizations providing realistic, accurate, and evaluable representations of the real-world environments are not yet available [40]. This paper describes the development of a visual-acoustic simulation integrating realistic acoustic emission and propagation into GIS-based 3D landscape visualizations to support landscape impact assessments. The visual simulation is generated employing a game engine. For the generation of synthetic wind turbine and environmental noise, the implementation of an emission synthesizer and a propagation filter as well as a reproduction system is developed. The visual and the acoustic simulations are then connected on the basis of linking parameters. These parameters can be controlled via the visual simulation and are transferred to the acoustic simulation to correctly auralize the corresponding soundscape. The focus of this article is the documentation of methods applied for developing the visual-acoustic simulation prototype.

## 4.2 METHODS

First, videos and sound recordings of a case study area had to be produced (Section 2.1). These recordings served as reference for the simulation of the landscape scenery including wind turbines as well as for the acoustic ambience of the wind turbine sound. Then, with regard to the visual simulation (Section 2.2) the essential processes are described: the coordinate transformation to import geodata, the calculation of the wind speed profile, and the visual optimizations to achieve an appropriate level of realism in the visualization. In the acoustic simulation (Section 2.3), the general simulation method and the developed calculation software are briefly described. Finally, Section 2.4 shows how both simulations are linked together.

### 4.2.1 REFERENCE VIDEO AND SOUND

The Mont Crosin in the Canton of Berne (Switzerland) was chosen as the reference site for the development of the visual-acoustic simulation tool. The wind farm comprises 16 wind turbines of the type Vestas [41]. On 11 October 2011 and 2 December 2011 the reference recordings were performed at Mont Crosin: video and sound recordings were produced as medium for comparison in order to develop the acoustic and visual simulation, and to test the validity of the integrated visual and acoustic simulation tool in a further step.

Prior to fieldwork, reference points at Mont Crosin were defined where videos and acoustic recordings were made, suitable to serve as reference data for both the acoustic and the visual simulation. Thereby, the fundamental auralization aspects were considered, consisting of emission (close vicinity of a wind turbine), ambiance (close to a forest edge), propagation (up to 500 m distance to a wind turbine), and multi-source recording (two wind turbines from different directions) [42]. Furthermore, constraints for the choice of viewpoints were given by the pedestrians' behavior, i.e., staying on the roads [43]. The contents of views were a frontal wind turbine, a wind turbine in the background, several wind turbines and no wind turbine. The viewing contents, except the "no wind turbine view", were required for comparing the human perception of the contents of the recordings and the simulations in the validation phase, carrying out a visual and acoustic landscape assessment.

In order to assess the acoustic impact of the wind turbines on the landscape under different wind conditions, a recording day with low wind (3.4 m/s at 10 m above ground) and one with strong wind (10.6 m/s at 10 m above ground) were chosen. The wind speed measurements were performed with a 3D ultrasonic anemometer. The main audio recordings were taken with a SPS200 Soundfield microphone, which, as a consequence of the strong wind, had to be mounted close to the ground and equipped with a Rycote Windshield and Windjammer. The visual recording system comprised a single-lens reflex camera with video recording capability. A 10–20 mm lens was used and fixed to 10 mm to capture a field of view of about 100 degrees. Videos were captured to catch the movement of the wind turbines and the vegetation influenced by the wind conditions on the recording days. The recording position in the field, viewing directions, wind speed, and wind direction parameters were compiled in a storyboard.

In addition, aerial images from the reference site were captured using an unmanned aerial vehicle (UAV). The aerial images were photogrammetrically processed, and a digital elevation model (DEM) as well as an orthophoto with a resolution in the decimeter range was calculated. This geodata can not only be used as an accurate and up-to date basis for landscape visualizations but also for determining the correct wind turbine positions in the landscape. The latter was particularly beneficial because the wind farm was recently enlarged and existing aerial images were outdated.

### 4.2.2 VISUAL SIMULATION

Using the Sandbox-Editor of Crytek's CryENGINE Version 3.3.9 [30], an interactive GIS-based 3D visualization with a high level of realism was generated. To visualize the landscape structure including the hilly terrain of the reference site, the visualization perimeter was set to a size of 8 × 8 km. Because the UAV data covered only 3 km² of the reference site of Mont Crosin, we decided to use the UAV data mainly for determining the correct placement of the 3D models. The 3D landscape simulation is therefore based on a DEM and an orthophoto of the Mont Crosin from swisstopo [44]. In addition, 3D models for vegetation, infrastructures, and wind turbines were added according to their actual locations, as shown in Figure 1.

A major task was to import and reference the geodata accurately because the game engine does not yet provide a direct GIS functionality. Therefore, conversion rules were defined, e.g., to import the digital elevation model so that the heights were displayed correctly in the virtual 3D model.

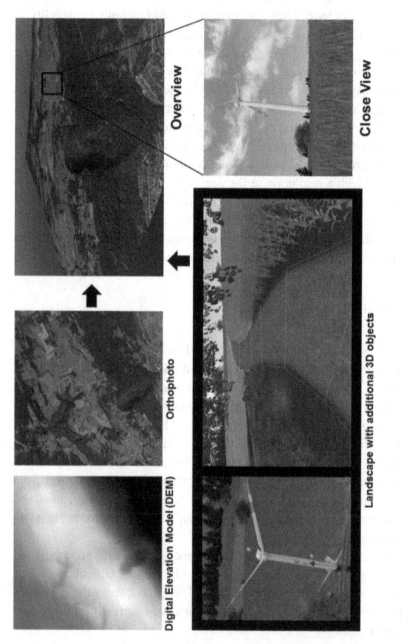

**FIGURE 1:** Visual simulation process of the wind farm at Mont Crosin.

**FIGURE 2:** Digital Elevation Model overlaid with an orthophoto and correctly placed wind turbines.

For this task, we took advantage from the CryENGINE's ability to interpret gray values from a height map [45]. Using an 8 bit height map allowing 256 gray values, an area with height differences over 255 m causes the height resolution to be larger than 1 m. In the reference region of Mont Crosin, the perimeter of 8 × 8 km has differences of about 900 m in height, including a mountain of about 1,000 m above sea level. Therefore, the height resolution is about 3.5 m per height unit (900 m/256 = 3.53 m). Working with 16 bit data files reduces this problem having over 65,000 grey values to represent even small height displacements. Before importing a greyscale terrain, e.g., a GIS-based DEM, the "Maximum Height" value in the engine has to be set. The "Maximum Height" corresponds to the maximum height difference in the perimeter that is the maximal height minus the minimal height of the DEM. The "Maximum Height" value sets the maximum terrain height which the imported terrain can be raised to [33]. Using the terrain editing functions in the engine, the elevation can be smoothed to achieve a better looking base model map [45]. The resulting basic model of the reference site is shown in Figure 2.

## 4.2.2.1 COORDINATE TRANSFORMATION

The auralization module requires information about the source and receiver position in order to link the respective sound to the view accurately. To get the current user position in the visualization in real world coordinates, transformation parameters have to be set which are applied in real-time to the CryENGINE coordinates. The auralization and the visualization are based on geodata in the Swiss coordinate system (LV03). Because the Swiss coordinate system is a Cartesian system and is defined to have only positive coordinate values within Switzerland, the conversion formula can be simplified to a translation. The translation parameter corresponds to the real world coordinates of the CryENGINE's origin, implicating the shift of the origin coordinates between the perimeter and the CryENGINE. By adding the translation parameters ($X_{CHofCEorigin}$/$Y_{CHofCEorigin}$) to the current position coordinates in CryENGINE ($X_{CEpos}$/$Y_{CEpos}$), the lateral real world position of the user in Swiss coordinates is known at any time:

$$X_{CHpos} = X_{CEpos} + X_{CHofCEorigin}$$

$$Y_{CHpos} = Y_{CEpos} + Y_{CHofCEorigin} \tag{1}$$

The height coordinate $Z_{CHpos}$ is dependent on the minimal height of the visualization perimeter in Swiss coordinates $Z_{CHmin}$ and the height value of the current position in CryENGINE $Z_{CEpos}$. The $Z_{CHmin}$ parameter corresponds to the 0 m height map value in CryENGINE.

$$Z_{CHpos} = Z_{CHmin} + Z_{CEpos} \tag{2}$$

Further deformation or distortion factors of the earth were not taken into account because of the small size of the perimeter, where these factors are negligible.

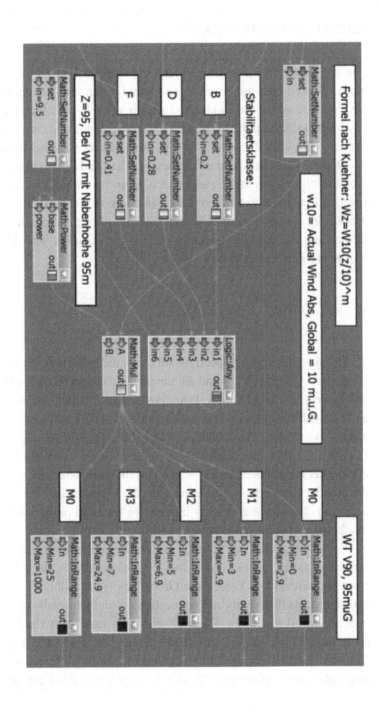

**FIGURE 3.** Section of the "Flow Graph" script developed in CryENGINE to control the specific parameters for wind speed profile calculation and corresponding wind turbine rotation in the virtual landscape model.

## 4.2.2.2 WIND SPEED PROFILE

The movement of the vegetation and the rotational speed of the turbine blades depend on wind direction and speed. Therefore, an algorithm was developed to adjust the wind speed and direction by user interaction (e.g., key inputs) affecting the movement of the vegetation and the wind turbine blades in the virtual model. This algorithm assumes a wind speed profile to calculate the wind speed at hub height. The wind field is determined by pressure differences in the atmosphere. Near the ground surface the wind speed is lower due to friction losses. In addition, the vertical wind speed profile depends on atmospheric stability caused by the temperature layers influencing the vertical movement of the air particles. The wind speed profile is calculated as follows [46,47,48]:

$$w_z = w_{10m}(z/10m)^m \tag{3}$$

To calculate the wind speed ($w_z$) at a specific height ($z$), the wind speed 10 m above ground ($w_{10m}$) and an atmospheric exponent ($m$) is needed. In this project, the wind speed 10 m above ground is defined as the global and adjustable wind speed parameter in CryENGINE influencing the vegetation movement. The wind speed at hub height is then calculated with Equation (3). The m-exponent is dependent on the atmospheric stability class. This is calculated based on the current solar radiation, respectively, the cloud cover at day or night time. In the visualization, this can be defined and changed interactively influencing the wind profile and therefore the theoretical wind speed at hub height.

The wind profile was scripted within the simulation using the visual scripting system "Flow Graph", which is embedded in the CryENGINE Sandbox editor [45]. The "Flow Graph" provides a library of functions and represents them and their connections visually to easily generate and control logical processes. Figure 3 shows an excerpt of the implemented formula Equation (3) in the "Flow Graph" script. As the wind turbine Vestas V90-2MW changes the rotational speed for different wind speed thresholds, rotation modes (M0-M3) were established in field (see Section

2.3). The wind speed at hub height indicates the rotation mode in which the wind turbine is currently operating. The output of Equation (3) (in the "Math:Mul"-Box in the middle of Figure 3) is directed to the wind turbine rotation modes (M0-M3), where the corresponding mode value regulates the speed function of the wind turbine rotor blades. In reality, the wind turbines never move synchronously. To run the wind turbines asynchronously in the simulation, a random value is added to the rotation speed and to the initial position of the rotor blades.

The script allows the user to control the wind speed and direction, allowing changing the wind turbine rotation in the engine in real-time while walking in the virtual 3D model.

### 4.2.2.3 VISUAL OPTIMIZATIONS

Landscape elements of 3D visualizations such as terrain, vegetation and built forms can have different levels of abstraction [49]. Each element has a geometric depiction (e.g., a polygon) with an association of textural properties on the geometric surface [49,50]. For their representation, Danahy [49], Lange [51] and Wissen [50] describe different scales in respect to their level of abstraction, ranging from abstract to realistic representations. Lange [51] states that from a modeling point-of-view, a visualization is more realistic, the more specific textures and geometries are used for modeling represented objects. Furthermore, a higher level of realism can be acquired by integrating and adjusting atmospheric conditions such as weather conditions, natural illumination or the simulation of dynamic processes [52,53,54]. In this project, we tried to visualize the reference locations in a high but appropriate level of realism. An appropriate level of realism should allow inducing perceptions of the simulated landscape similar to the reference recordings. To achieve this, the following visual optimizations were applied.

The reference videos were analyzed with regard to the wind turbine orientation and rotation, the lighting and weather conditions, and the vegetation movement as well as the structure of the vegetation in the landscape (type, height, density and distribution). The vegetation was animated according to the measured wind speed in field and the movement in the reference video, and with the additional help of the aerial images the vegetation was placed accurately in the visualization.

The lighting conditions were adjusted using the CryENGINE's ability to simulate different daytimes. Based on the reference data collected in field, the daytime and the weather conditions, such as cloud cover, wind speed, wind direction, sun intensity, and sun settings were set in the game engine. The sun settings allow adjusting the position of the sun, the sun direction (the direction from where the sun rises), and the daytime. The sun position implies the distance the perimeter is located from the North Pole to the South Pole, and can be estimated using the real latitude position of the perimeter (for the reference site: 48°N in WGS84). As the CryENGINE sun position values range from 0 (=North Pole) to 180 (=South Pole) [33], the sun position value in the sun settings is 42 for the reference site.

Furthermore, the orthophoto color was adjusted to match with the current ground color in the reference videos, using an image editing application. This was necessary for the strong wind situation because the orthophoto was captured at a slightly different season and weather condition than the reference videos.

For the strong wind condition, the sky was covered with fast moving clouds. As these moving clouds are giving an impression of a typical windy environment, this effect has to be implemented as well in the visual simulation, using CryENGINE's 3D clouds [33]. Corresponding to the cloud moving direction in the reference video, the moving path for the major clouds was defined in the visual simulation, leading to an appropriate visual windy effect in the landscape visualization.

For the low wind situation, direct sun exposure was present on the recording day, generating fast moving shadows on the ground from the wind turbine rotors as well as smaller, stationary shadows from the vegetation. Due to the dynamic shadow calculation of the CryENGINE the shadows could be reconstructed accurately with respect to the reference recordings.

The available 3D wind turbine object models obtained from the software library of WindPRO [55] were resized in the software 3D Studio Max from Autodesk [56] corresponding to the wind turbine sizes on the reference site, including the wind turbines' heights, mast diameters, and the rotor blade dimensions. Additionally, the wind turbine color material was adjusted in order to match the appearances and visual attributes of the real ones. To obtain a 3D wind turbine object file, readable in CryENGINE (e.g., .cgf, .mtl and .dds), the exporter of the CryENGINE 3ds Max Plugin [57] was used. The wind turbine rotor and the mast were imported as separate objects to animate the rotor independently from the mast. In the CryENGINE's "Flow Graph", the rotor is linked to a rotation function, which specifies the movement of the rotation direction (x,y,z) and the rotational speed. Linking the wind turbine rotor to the mast using CryENGINE's "Link Object" function, it is possible to change the entire wind turbine orientation based on the wind direction, while all wind turbine rotors can be steered individually with respect to the wind speed.

All these visual adjustments led to an appropriate landscape visualization providing a level of realism suitable for a visual perception evaluation of the simulated landscapes in the validation part of the project [40].

### 4.2.3 ACOUSTIC SIMULATION

The acoustic simulation consists of the generation of the wind turbine emission audio signal (1), the filtering to account for the frequency dependent sound propagation attenuation (2), the generation of a natural background audio signal (3), and the reproduction by loudspeakers (4) as shown in the block diagram in Figure 4.

In a first step, the emission synthesizer (1) was developed to generate the emission audio signal. A key issue was the investigation of the emission signal dependency from turbine type and operation condition, wind speed, and temperature stratification. Hence, in addition to the field survey for the reference video and sound from October to December 2011 (see Section 2.1) several emission recordings of wind turbines at varying operation conditions were taken at the reference site at Mont Crosin. Instead of an expected continuous relation between wind speed and rotational speed, the measurements on Vestas V90-2MW turbines revealed discrete operational modes (M0–M3). A synthesis model for the generation of emission audio signals of the Vestas V90-2MW was developed and implemented in the software Matlab [58].

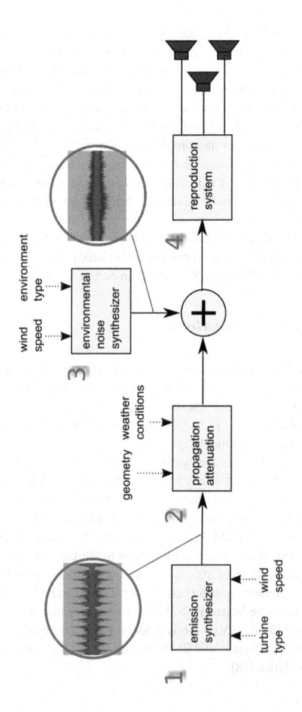

**FIGURE 4:** Block diagram for the generation of synthetic wind turbine and environmental noise.

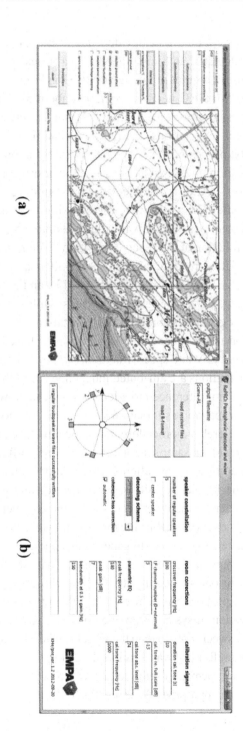

**FIGURE 5:** Graphical user interfaces of the computer programs (a) AuraPRO and (b) RePRO.

In a second step, the frequency dependent sound propagation attenuation (2) was calculated and implemented as a series of digital filters. For that purpose, the computer program AuraPRO in Figure 5a was developed [38]. As input, AuraPRO needs information about the linking parameters from the visual simulation (see Section 2.4), such as the coordinates of the source and receiver positions, the weather conditions, and the emission audio file, to be processed. The topography information is obtained from the digital elevation model, which was also the basis for the visual simulation (see Figure 1). Additional objects such as buildings can be defined in a separate text file. The audio wave file containing the emission signal is read and processed to an output wave file. The output represents the sound pressure time function at the receiver position. So far, the propagation includes geometrical spreading, air absorption, attenuation by obstacles, attenuation by foliage (all according to ISO 9613), and ground effects.

In a third step, possible vegetation noise is synthesized (3). The model considers the vegetation geometry, the vegetation type, and wind speed [38]. Vegetation noise is modulated in amplitude based on a turbulence model that predicts wind speed fluctuations. As a simplification, these wind speed fluctuations are attributed independently of the individual vegetation cells and not referenced to a location dependent wind speed field.

The sound propagation simulation allows a receiver to change the position over time. Hereby, a procedure had to be developed to map emission samples onto receiver samples for arbitrary time-shifts without audible artifacts. This non-linear operation correctly models the Doppler Effect, that is to say the frequency shift between emitted and received signal in case of a relative movement between source and receiver.

The last step covers a suitable mapping of the synthesized signals to a system of loudspeakers (4) in order to generate an appropriate listening impression regarding sound pressure levels and directional information. In our application, an ambisonics rendering strategy was chosen. Hereby, an arbitrary number of loudspeakers can be used and different coding strategies allow for the optimization of system properties such as localization accuracy and extension of the sweet spot. To map the signals to the loudspeaker system, the computer program RePRO was developed, see Figure 5b. Five loudspeakers were arranged in a pentagon-setup to generate an optimal listening impression that allows an appropriate determination of the sound source direction.

**FIGURE 6:** Overlaid head-up display (HUD) information of the current position in field and viewing angle (upper left corner) and the current wind speed and direction (bottom left corner).

### 4.2.4 LINKING ACOUSTIC TO VISUAL SIMULATION

A concept was developed to connect the acoustic simulation output adequately to the 3D landscape model. As it is not yet possible to generate the acoustic simulation in real-time, the audio files have to be linked to the visualizations in a post-process. This includes three tasks:

1.  Rendering images for a video out of the CryENGINE from a viewpoint or a walk path.
2.  Saving all relevant parameters which are needed to calculate the audio files (Section 2.3) correctly into a file and providing them to the acoustic simulation calculation.
3.  Converting the images into a video, and linking the synthesized audio files to the video using the software Adobe Premiere Pro [59].

In the first task, images are recorded simultaneously to the movement through the landscape, using a capturing function in the CryENGINE's

"Flow Graph" script. In the second task, the parameters relevant for both the acoustic and the visual simulation have to be streamed in real-time into a single parameter file at the same time the images are recorded (step 2). These parameters are: the position, the viewing angle, the wind turbine positions, the turbine rotational speed, the initial rotor position, the wind direction, and the wind speed. Nakevska et al. [29] established an interaction with the game engine generating XML files, where e.g., each object is a child with attributes of the position and the rotation. In our case, the relevant parameters are gathered in the visualization software by the developed "Flow Graph" script (see Section 2.2.2) and streamed into an external XML-file as well. The developed approach allows accessing these parameters in the virtual landscape model and provides the data in a suitable format as input for the auralization model. Based on these parameters, the acoustic simulation can be calculated. In the third task, the challenge was to correctly synchronize the movement of the animated wind turbine to the sound files. Therefore, a function was developed in the "Flow Graph" script. First, the actual parameter values are shown in an overlay, the so-called head-up display (HUD) to control the settings of the relevant parameters (see Figure 6).

Then, with a pre-defined key input in the "Flow Graph" script, the HUD parameter information can be hidden and a START sign appears simultaneously for 0.3 s in the HUD, initiating the rotation of the wind turbines. This START sign indicates the starting time of the simulation and of the parameter file generation. By producing the simulation videos in Adobe Premiere Pro, the corresponding acoustic files can be correctly applied to the visualization based on the visual START information.

## 4.3 RESULTS

Implementing the presented method, we generated correctly linked visual-acoustic videos of the simulated landscapes. Videos were produced for all the reference locations of Mont Crosin. Figure 7 shows three out of five reference locations, on the left side the reference locations captured in field (see Section 2.1) and on the right side the correspondingly simulated locations.

*Reference landscapes*      *Simulated landscapes*

**FIGURE 7:** Three out of five reference (Left) and simulated (Right) landscape videos.

## 4.4 DISCUSSION

We presented an approach to work with GIS-data in a game engine, to simulate wind turbine noise correctly, and to link the noise simulation to the virtual landscape. This method demonstrates an appropriate way for generating detailed real world landscape representations and for linking the auralization to the virtual landscape.

We produced a GIS-based virtual landscape model employing Crytek's CryENGINE 3. The use of this game engine offered a huge advantage for the visual-acoustic simulation. As Crytek's CryENGINE includes a high performance physics engine, it was possible to create visualizations with great amounts of detail, including vegetation movement, lighting conditions, shadow calculation, and 3D cloud generation. Furthermore, CryENGINE's "Flow Graph" allowed scripting logical processes in an interactive manner, where all relevant parameters for the visual and acoustic simulations can be modified by user interaction and exported in real-time. These features allow linking the visual and acoustic simulation in an appropriate way.

However, since the visualization software is not connected directly to a geographical information system (GIS), it was a major task to integrate and to represent the spatial data accurately and coherently in the game engine. An appropriate transformation of a specific real world coordinate system into internal game engine coordinates is, thus, crucial. The documentation of the straightforward translation approach in this article provides guidance for generating fast and simple GIS-based visualizations with CryENGINE. The translation allows not only for importing geodata and real world coordinate based 3D objects, but also for exporting the linking parameters required for the acoustic simulation adequately.

With the help of the developed tool, it is possible to link the acoustics for an arbitrary wind farm design to a GIS-based landscape visualization in the game engine. Linking parameters defined by predefined settings and changed by user interaction in the visual simulation are stored in a parameter file available for calculating the correct auralization. The wind farm design can be changed interactively and subsequently, the sound is calculated accordingly for the new setting. The user can, thus, choose any location in the virtual landscape and the auralization of the wind turbines can be calculated based on the actual parameters of the location.

The innovative aspects of the developed auralization computer programs are the development of new algorithms and strategies for simulating spatially explicit sound of wind turbines taking into account the environmental context. However, the audio synthesizer is not yet able to calculate and reproduce the audio in real-time. Therefore, the sound files have to be calculated offline in a post-processing step according to the parameter file exported from the visualization module. Currently, the audio information is rendered to five loudspeakers arranged in a pentagon, allowing for optimized source localization with sufficient angular resolution in all directions. In a subsequent step, the audio files are manually linked to the animated virtual landscape video. To avoid the manual integration of video and audio, it would be preferable to use batch procedures that automatically perform the combination (muxing) of video and audio.

From a technical perspective, an important improvement of the prototype would be the development of real-time functionality of the simulation tool. As it is already possible to change and store the actual linking parameters in real-time, investigations have to be made to generate and auralize the acoustics at the same time the user moves individually through the 3D landscape.

As a next step, this prototype of a visual-simulation tool has to be validated. Both Bishop [19] and Lange [8] state that further developments in visualization techniques help people to understand future landscape changes and therefore influence the decision making process, but it has to be considered how landscape simulations are perceived by humans. Therefore, a validation is needed, where a comparative evaluation is conducted, comparing videos of the simulated wind farm to videos of the real wind farm. The validated tool can then be implemented for further assessments of wind farm scenarios, e.g., in participatory workshop settings as Bishop [19] suggests, or in an acceptability study to assess the impact of different wind farm designs in different landscape contexts. In this way, the development of recommendations for wind farm planning and for exploiting socially-accepted wind energy locations can be supported.

## 4.5 CONCLUSIONS

We developed a prototype of a GIS-based visual-acoustic simulation tool suitable for assessing and discussing choices of locations for wind turbines and their impact on perceived landscape quality in participatory processes. The employed game engine Crytek's CryENGINE 3 has delivered a suitable software program, which proved successful for both (1) realistic 3D visualization of animated wind farms with high level of realism based on

GIS-data, as well as (2) providing input parameters for linking spatially explicit acoustic simulations to the accurate locations in the virtual model. With the documentation of the method, we contribute to establishing guidance to generate such visual-acoustic simulations. Overall, this prototype can contribute significantly to enhancing the available set of instruments for conscious landscape development with wind farms. The integration of spatial sound simulation into the correlating virtual landscapes can allow for an improved impact assessment and, thus, it may provide a better, more comprehensible decision basis than conventional tools, such as decibel maps or photomontages.

## REFERENCES

1. Wolsink, M. Wind power implementation: The nature of public attitudes: Equity and fairness instead of "backyard motives". Renew. Sustain. Energy Rev. 2007, 11, 1188–1207, doi:10.1016/j.rser.2005.10.005.
2. Devine-Wright, P. Beyond NIMBYism: Towards an integrated framework for understanding public perceptions of wind energy. Wind Energy 2005, 8, 125–139, doi:10.1002/we.124.
3. Cowell, R. Wind power, landscape and strategic, spatial planning—The construction of "acceptable locations" in Wales. Land Use Policy 2010, 27, 222–232.
4. Bundesamt für Energie BFE; Bundesamt für Umwelt, Wald und Landschaft BUWAL; Bundesamt für Raumentwicklung ARE. Konzept Windenergie Schweiz—Grundlagen für die Standortwahl von Windparks; Bundesamt für Energie BFE, Bundesamt für Umwelt, Wald und Landschaft BUWAL, Bundesamt für Raumentwicklung ARE: Bern, Switzerland, 2004.
5. Hall, N.; Ashworth, P.; Devine-Wright, P. Societal acceptance of wind farms: Analysis of four common themes across Australian case studies. Energy Policy 2013, 58, 200–208, doi:10.1016/j.enpol.2013.03.009.
6. Pepermans, Y.; Loots, I. Wind farm struggles in Flanders fields: A sociological perspective. Energy Policy 2013, 59, 321–328, doi:10.1016/j.enpol.2013.03.044.
7. Bundesamt für Energie BFE. Energieforschung. In Code of Conduct für Windkraftprojekte, Machbarkeitsstudie—Schlussbericht; Swiss Federal Office for Energy: Bern, Switzerland, 2009.
8. Lange, E. 99 volumes later: We can visualise. Now what? Landsc. Urban Plan. 2011, 100, 403–406, doi:10.1016/j.landurbplan.2011.02.016.
9. Shepherd, D.; McBride, D.; Welch, D.; Hill, E.; Dirks, K. Evaluating the impact of wind turbine noise on health-related quality of life. Noise Health 2011, 13, 333–339, doi:10.4103/1463-1741.85502.
10. Mijuk, G. Der Wind dreh. In NZZ am Sonntag; Neue Zürcher Zeitung AG: Zurich, Switzerland, 2010; pp. 22–23.

11. Pedersen, E.; Larsman, P. The impact of visual factors on noise annoyance among people living in the vicinity of wind turbines. J. Environ. Psychol. 2008, 28, 379–389, doi:10.1016/j.jenvp.2008.02.009.

12. Laubwerk GmbH. 3D Plants for CG Artists. Available online: http://www.laubwerk. com (accessed on 29 October 2013).

13. Xfrog Inc. 3D Trees and 3D Plants for CG Artists. Available online: http://xfrog.com (accessed on 29 October 2013).

14. IDV Inc. SpeedTree Animated Trees & Plants Modeling & Render Software. Available online: http://www.speedtree.com (accessed on 29 October 2013).

15. Ghadirian, P.; Bishop, I.D. Integration of augmented reality and GIS: A new approach to realistic landscape visualisation. Landsc. Urban Plan. 2008, 86, 226–232, doi:10.1016/j.landurbplan.2008.03.004.

16. Stock, C.; Bishop, I.D. Linking GIS with real-time visualisation for exploration of landscape changes in rural community workshops. Virtual Real. 2006, 9, 260–270, doi:10.1007/s10055-006-0023-9.

17. Stock, C.; Bishop, I.D.; Green, R. Exploring landscape changes using an envisioning system in rural community workshops. Landsc. Urban Plan. 2007, 79, 229–239, doi:10.1016/j.landurbplan.2006.02.010.

18. Stock, C.; Bishop, I.D.; O'Connor, A. Generating Virtual Environments by Linking Spatial Data Processing with a Gaming Engine. In Trends in Real-time Landscape Visualization and Participation; Buhmann, E., Paar, P., Bishop, I.D., Lange, E., Eds.; Wichmann: Heidelberg, Germany, 2005; pp. 324–329.

19. Bishop, I.D. Landscape planning is not a game: Should it be? Landsc. Urban Plan. 2011, 100, 390–392, doi:10.1016/j.landurbplan.2011.01.003.

20. Lange, E.; Hehl-Lange, S. Combining a participatory planning approach with a virtual landscape model for the siting of wind turbines. J. Environ. Plan. Manag. 2005, 48, 833–852, doi:10.1080/09640560500294277.

21. Otero, C.; Manchado, C.; Arias, R.; Bruschi, V.M.; Gómez-Jáuregui, V.; Cendrero, A. Wind energy development in Cantabria, Spain. Methodological approach, environmental, technological and social issues. Renew. Energy 2012, 40, 137–149, doi:10.1016/j.renene.2011.09.008.

22. Friese, K.-I.; Herrlich, M.; Wolter, F.-E. Using Game Engines for Visualization in Scientific Applications. In New Frontiers for Entertainment Computing; Ciancarini, P., Nakatsu, R., Rauterberg, M., Roccetti, M., Eds.; Springer: Boston, MA, USA, 2008; Volume 279, pp. 11–22.

23. Lewis, M.; Jacobson, J. Game engines in scientific research. Commun. ACM 2002, 45, 27–31.

24. Herwig, A.; Paar, P. Game Engines: Tools for Landscape Visualization and Planning? In Trends in GIS and Virtualization in Environmental Planning and Design; Buhmann, E., Nothelfer, U., Pietsch, M., Eds.; Wichmann: Heidelberg, Germany, 2002; pp. 161–172.

25. Germanchis, T.; Pettit, C.; Cartwright, W. Building a three-dimensional geospatial virtual environment on computer gaming technology. J. Spat. Sci. 2004, 49, 89–95, doi:10.1080/14498596.2004.9635008.

26. GarageGames. Game Development Tools and Software. Available online: http:// www.garagegames.com (accessed on 29 October 2013).

27. Jacobson, J.; Lewis, M. Game engine virtual reality with CaveUT. Computer 2005, 38, 79–82, doi:10.1109/MC.2005.126.
28. Epic Games Inc. Game Engine Technology by Unreal. Available online: http://www.unrealengine.com (accessed on 29 October 2013).
29. Nakevska, M.; Vos, C.; Juarez, A.; Hu, J.; Langereis, G.; Rauterberg, M. Using Game Engines in Mixed Reality Installations. In Entertainment Computing—ICEC 2011; Anacleto, J., Fels, S., Graham, N., Kapralos, B., Saif El-Nasr, M., Stanley, K., Eds.; Springer: Berlin/Heidelberg, Germany, 2011; Volume 6972, pp. 456–459.
30. Crytek GmbH. MyCryENGINE. Available online: http://mycryengine.com (accessed on 29 October 2013).
31. Germanchis, T.; Cartwright, W.; Pettit, C. Using Computer Gaming Technology to Explore Human Wayfinding and Navigation Abilities within the Built Environment. In Proceedings of the XXII International Cartographic Conference, A Coruña, Spain, 9–16 July 2005; pp. 11–16.
32. Germanchis, T.; Cartwright, W.; Pettit, C. Virtual Queenscliff: A Computer Game Approach for Depicting Geography. In Multimedia Cartography; Cartwright, W., Peterson, M., Gartner, G., Eds.; Springer: Berlin/Heidelberg, Germany, 2007; pp. 359–368.
33. Tracy, D.; Tracy, S. CryENGINE 3 Cookbook: Over 90 Recipes Written by Crytek Developers for Creating Third-Generation Real-Time Games; Packt Publishing Ltd.: Birmingham, UK, 2011.
34. Herrlich, M. A Tool for Landscape Architecture Based on Computer Game Technology. In Proceedings of the 17th International Conference Artificial Reality and Telexistence, Esbjerg, Denmark, 28–30 November 2007; pp. 264–268.
35. Herrlich, M.; Holle, H.; Malaka, R. Integration of CityGML and Collada for High-Quality Geographic Data Visualization on the PC and Xbox 360. In Entertainment Computing—ICEC 2010; Yang, H., Malaka, R., Hoshino, J., Han, J., Eds.; Springer: Berlin/Heidelberg, Germany, 2010; Volume 6243, pp. 270–277.
36. Oerlemans, S.; Sijtsma, P.; Méndez López, B. Location and quantification of noise sources on a wind turbine. J. Sound Vib. 2007, 299, 869–883, doi:10.1016/j.jsv.2006.07.032.
37. Schepers, J.G.; Curvers, A.; Oerlemans, S.; Braun, K.; Lutz, T.; Herrig, A.; Wuerz, W.; Mantesanz, A.; Garcillan, L.; Fischer, M.; et al. SIROCCO: Silent Rotors by Acoustic Optimisation. In Proceedings of the 2nd International Meeting on Wind Turbine Noise, Lyon, France, 20–21 September 2007.
38. Heutschi, K.; Pieren, R.; Müller, M.; Manyoky, M.; Wissen Hayek, U.; Eggenschwiler, K. Auralization of wind turbine noise: Propagation filtering and vegetation noise synthesis. Acta Acust. United Acust. 2014, 100, 13–24, doi:10.3813/AAA.918682.
39. Heutschi, K.; Pieren, R. Auralization of Wind Turbines. In Proceedings of the AIA-DAGA 2013 Conference on Acoustics, Merano, Italy, 18–21 March 2013.
40. Manyoky, M.; Hayek, U.W.; Klein, T.M.; Pieren, R.; Heutschi, K.; Grêt-Regamey, A. Concept for Collaborative Design of Wind Farms Facilitated by an Interactive GIS-Based Visual-Acoustic 3D Simulation. In Peer Reviewed Proceedings of Digital Landscape Architecture 2012 at Anhalt University of Applied Sciences; Buhmann, E., Ervin, S., Pietsch, M., Eds.; Wichmann: Heidelberg, Germany, 2012; pp. 297–306.

41. Vestas Wind Systems A/S. Vestas Wind Systems. Available online: http://www.vestas.com (accessed on 29 October 2013).
42. Vorländer, M. Auralization: Fundamentals of Acoustics, Modelling, Simulation, Algorithms and Acoustic Virtual Reality; Springer: Berlin/Heidelberg, Germany, 2008.
43. Braun, S.; Ziegler, S. Windlandschaft: Neue Landschaften mit Windenergieanlagen; Wissenschaftlicher Verlag Berlin (WVB): Berlin, Germany, 2006.
44. Federal Office of Topography Swisstopo. Swisstopo—Knowing Where. Available online: http://www.swisstopo.ch (accessed on 29 October 2013).
45. Tracy, S.; Reindell, P. CryENGINE 3 Game Development: Beginner's Guide; Packt Publishing Ltd.: Birmingham, UK, 2012.
46. Kühner, D. Excess attenuation due to meteorological influences and ground impedance. Acta Acust. United Acust. 1998, 84, 870–883.
47. Van den Berg, G.P. Do Wind Turbines Produce Significant Low Frequency Sound Levels? In Proceedings of the 11th International Meeting on Low Frequency Noise and Vibration and its Control, Maastricht, The Netherlands, 30 August–1 September 2004; Volume 13, pp. 367–376.
48. Heutschi, K.; Eggenschwiler, K. Lärmermittlung und Massnahmen zur Emissionsbegrenzung bei Windkraftanlagen. Untersuchungsbericht Nr. 452'460, int. 562.2432; Eidg. Materialprüfungs- und Forschungsanstalt, Abteilung Akustik: Dübendorf, Switzerland, 2010.
49. Danahy, J. A Set of Visualization Data Needs in Urban Environmental Planning & Design for Photogrammetric Data. In Automatic Extraction of Man-Made Objects from Aerial and Space Images (II); Gruen, A., Baltsavias, E., Henricsson, O., Eds.; Birkhäuser: Basel, Switzerland, 1997; pp. 357–366.
50. Wissen, U. Virtuelle Landschaften zur Partizipativen Planung: Optimierung von 3D-Landschaftsvisualisierungen zur Informationsvermittlung; vdf Hochschulverlag AG: Zürich, Switzerland, 2009.
51. Lange, E. Issues and Questions for Research in Communicating with the Public through Visualizations. In Trends in Real-Time Landscape Visualization and Participation; Buhmann, E., Paar, P., Bishop, I.D., Lange, E., Eds.; Wichmann: Heidelberg, Germany, 2005; pp. 16–26.
52. Paar, P.; Schroth, O.; Lange, E.; Wissen, U.; Schmid, W.A. Steckt der Teufel im Detail? Eignung Unterschiedlicher Detailgrade von 3D-Landschaftsvisualisierung für Bürgerbeteiligung und Entscheidungsunterstützung. In Proceedings of the 9th International Conference on Urban Planning and Regional Development in the Information Society, Vienna, Austria, 25–27 February 2004; pp. 535–541.
53. Achleitner, E.; Schmidinger, E.; Voigt, A. Dimensionen Eines Digitalen Stadtmodelles am Beispiel Linz. In Proceedings of the 8th International Conference on Urban Planning and Regional Development in the Information Society, Vienna, Austria, 25 February–1 March 2003; pp. 171–179.
54. Erwin, S.M. Digital landscape modeling and visualization: A research agenda. Landsc. Urban Plan. 2001, 54, 49–62, doi:10.1016/S0169-2046(01)00125-6.
55. EMD International A/S. Danish Software and Consultancy Firm Assisting Countries with Planning and Documenting Environmental Energy Projects. Available online: http://www.emd.dk (accessed on 29 October 2013).

56. Autodesk Inc. 3D Design, Engineering & Entertainment Software. Available online: http://www.autodesk.com (accessed on 29 October 2013).
57. Crytek GmbH. Installing the 3ds Max Tools—Doc 3. Asset Creation Guide. Available online: http://freesdk.crydev.net/display/SDKDOC3/Installing+the+3ds+Max+Tools (accessed on 30 October 2013).
58. Pieren, R.; Heutschi, K.; Müller, M.; Manyoky, M.; Eggenschwiler, K. Auralization of wind turbine noise: Emission synthesis. Acta Acust. united Acust. 2014, 100, 25–33, doi:10.3813/AAA.918683.
59. Adobe Systems Inc. Adobe Premiere Pro CC. Available online: http://www.adobe.com/PremierePro (accessed on 29 October 2013).

# PART II

# SOCIAL, ECONOMIC, AND ENVIRONMENTAL ISSUES

## CHAPTER 5

# Assessment of Wind Power Generation Along the Coast of Ghana

MUYIWA S. ADARAMOLA, MARTIN AGELIN-CHAAB, AND SAMUEL S. PAUL

## 5.1 INTRODUCTION

Global decline in fossil fuel reserves, damaging effects of global warming, and rising energy demand due to increasing population have necessitated the need for more research and development of low-carbon sources of energy. There is consensus that wind energy is the leading alternative energy source and the fastest growing segment of the global renewable energy industry [1] and [2]. Investments in wind energy have been growing at 22% average for the past 10 years, with wind energy constituting 2.5% of the global electricity supply and projected to rise to 8–12% by 2020 [1]. Ghana is not considered a player in the wind energy sector and does not, in fact, currently produce any significant amount of wind power. Nevertheless, a conservative estimate suggests that over 1000 km$^2$ of land area exist with moderate-to-excellent wind resource potential [3].

Ghana depended solely on hydro power from the mid-60s to the late 90s. However, the country added thermal power in the late 90s due to

*Reprinted from* Energy Conversion and Management, *77, Adaramola MS, Agelin-Chaab M, Paul SS, Assessment of Wind Power Generation Along the Coast of Ghana, Pages 61–69, Copyright 2014, with permission from Elsevier.*

increased population and poor rainfall patterns. The country's electricity supply is primarily from hydro and thermal power plants. According the Ghana's Energy Commission, the total installed electricity generation capacity in 2012 was 2280 MW [4]. This is made up of 1180 MW from hydro (51.8%) and 1100 MW from thermal (48.2%). In the same year the electricity consumed was 9258 GW h, out of which 2931 GW h constituted residential use (i.e., ~32%). Unfortunately, there still remains close to 28% of households that do not have access to electricity in the country [5]. As a result of cheap natural gas from Nigeria and Ghana's newly discovered oil fields, Ghana's electricity supply mix is likely to be dominated by thermal in the near future. The country is beginning to add renewable energy to its energy portfolio. The Volta River Authority, the main generator of electricity in Ghana has just completed a 2 MW solar power plant, which is the largest in mainland West Africa [6]. The company is also projected to generate additional 150 MW from wind power and 14 MW from solar energy by 2015 [6]. As the country diversifies its energy sources to meet its rapidly growing energy demand, due to the expanding economy and growing population, it should consider adding more wind energy. This is because advances in wind turbine technology are pushing down the cost of wind energy so quickly that it will be cost competitive with new thermal plants in the near future.

Wind data in Ghana is traditionally measured by the country's Meteorological Services Department (MSD). It conducts measurements at its 22 synoptic stations across the country. The problem with MSD's data is that the wind speeds are obtained at only 2 m (above ground level, a.g.l.). To obtain more suitable wind data for wind energy application, Ghana's Energy Commission (EC) conducted its own wind data measurement along the coastal areas of the country at 11 stations in the late 1990s. The EC's measurements were obtained at a height of 12 m (a.g.l.). In addition, the EC conducted new simultaneous measurements at both 2 m and 12 m (a.g.l.) at some of MSD's locations, and used comparative studies and statistical techniques to extrapolate all wind data at the MSD's 22 stations to 12 m and 50 m (a.g.l.) [7]. All these wind data measurement activities are well documented in a number of studies [7], [8] and [9]. The data indicated that the coastal region of Ghana has the most promising wind energy potential in the country. Nonetheless, extensive analyses have not been performed on the data in order to provide energy policy makers diverse view points for better informed decision making.

The objective of this article is therefore to evaluate the wind energy potential in six selected locations (Adafoah, Anloga, Aplaku, Mankoadze, Oshiyie and Warabeba). These are all located along the coast of Ghana as indicated approximately on the map in Fig. 1 and Table 1. The study will use the existing data in [7] to perform the economic analysis on selected small to medium size commercial wind turbines. This information will help the government of the country and other stakeholders make better informed decisions regarding investment in wind energy resources.

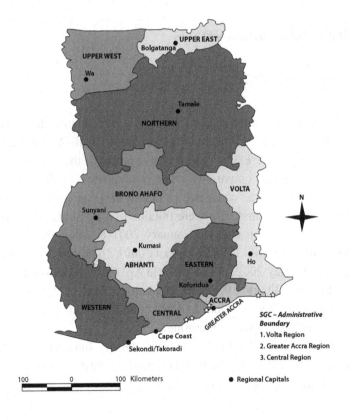

**FIGURE 1:** Regional map of Ghana with star symbols indicating the approximate wind measurement locations selected for this study. From East to West: Anloga, Adafoah, Aplaku, Oshiyie, Warabeba and Mankoadze. (Source of map: http://www.adraghana.org/).

**TABLE 1:** Geographical coordinates of the selected locations.

| Site | Latitude (°) | Longitude (°) | Elevation[a] (m) |
|------|------|------|------|
| Adafoah | 5.78N | 0.63E | 2.0 |
| Anloga | 5.80N | 0.90E | 5.0 |
| Aplaku | 5.51N | 0.32W | 0.5 |
| Mankoadze | 5.32N | 0.67W | 7.0 |
| Oshiyie | 5.50N | 0.35W | 1.0 |
| Warabeba | 5.33N | 0.62W | 5.0 |

*[a]Elevation above sea level.*

## 5.2 WIND SPEED DATA ANALYSIS

### 5.2.1 METHODOLOGY

The wind speed data for the selected locations were obtained from the EC. Wind speed was captured at the height 12 m (a.g.l.) in all the selected sites, except for Anloga where wind speed was captured at the height of 20 m. In order to reasonably compare wind speed parameters in all the sites, the wind data at Anloga was converted to height of 12 m (using power law expression). The acquired data were obtained on an hourly basis over a period of 12 months (in Oshiyie, January 2001 to December 2001), 13 months (in Adafoah, July 1999 to August 2000), 16 months (in Warabeba, January 2001 to May 2002), 18 months (in Aplaku and Mankoadze, January 2001 to July 2002) to 19 months (in Anloga, May 2006 to December 2007). From these hourly wind data, the monthly wind speed and other wind speed parameters were determined.

The two-parameter Weibull probability density function was used to analyze the wind data. The Weibull probability density function is given [10], [11], [12], [13] and [14] as:

$$f(v) = \left(\frac{k}{c}\right)\left(\frac{v}{c}\right)^{k-1} \exp\left[-\left(\frac{v}{c}\right)^{k}\right] \tag{1}$$

where f(v) is the probability of observing wind speed (v), k is dimensionless Weibull shape parameter, and c is the Weibull scale parameter (m/s). The corresponding cumulative distribution F(V) is the integral of the probability density function and it is expressed as:

$$F(v) = 1 - \exp\left[-\left(\frac{v}{c}\right)^k\right] \tag{2}$$

The Weibull parameters were calculated using standard deviation method among other methods listed in [15]. This method is useful due to the following reasons [12], [16], [17] and [18]: appropriateness in situations, where only the mean wind speed and standard deviation are available; it gives better results than graphical method and has a relatively simple expression when compared with other methods. When the value of k is set to two, the above expressions become one-parameter Rayleigh distribution function and are expressed mathematically, respectively, as:

$$f(v) = \frac{\pi}{2}\left(\frac{v}{v_m^2}\right)\exp\left[-\frac{\pi}{4}\left(\frac{v}{v_m}\right)^2\right] \tag{3}$$

$$F(v) = 1 - \exp\left[-\frac{\pi}{4}\left(\frac{v}{v_m}\right)^2\right] \tag{4}$$

where $v_m$ is the mean wind speed and is defined mathematically as:

$$v_m = \frac{1}{n}\sum_{i=1}^{n} v_i \tag{5}$$

The Rayleigh distribution function is generally used in a situation where only the mean wind speeds data are available for the desired location [19]. The Weibull parameter factor k and the Weibull scale parameter can be estimated by the mean wind speed-standard deviation method using the following expressions [12] and [20]:

$$k = \left(\frac{\sigma}{v_m}\right)^{-1.086} \quad 1 \leq k \leq 10 \tag{6}$$

and,

$$c = \frac{v_m}{\Gamma\left(1 + \frac{1}{k}\right)} \tag{7}$$

where $\sigma$ is the standard deviation and $\Gamma$ is the gamma function. Instead of using Eq. (7) above, a scale factor can be determined from the following expressions given by [21]:

$$c = \frac{v_m k^{2.6674}}{0.184 + 0.816 k^{2.73855}} \tag{8}$$

The standard deviation and the gamma function are respectively expressed as [16]:

$$\sigma = \left[\frac{1}{n-1}\sum_{i=1}^{n}(v_i - v_m)^2\right]^2 \tag{9}$$

$$\Gamma(x) = \int_0^{\infty} t^{x-1} e^{-t} dt \tag{10}$$

where n is the number of data set.

In addition to the mean wind speed of a site, the other useful characteristics wind speeds parameters are the most probable wind speed ($v_f$), and the wind speed carrying maximum energy ($v_e$). They can be estimated from the following expressions for Weibull distribution function:

$$v_f = c \left( \frac{k-1}{k} \right)^{1/k} \tag{11}$$

$$v_e = c \left( \frac{k+2}{k} \right)^{1/k} \tag{12}$$

For the Rayleigh distribution function, these expressions reduce to [19] and [22]:

$$v_f = v_m \sqrt{\frac{2}{\pi}} \tag{13}$$

$$v_e = 2v_m \sqrt{\frac{2}{\pi}} \tag{14}$$

The significance of the wind speed carrying maximum energy is that it is closely related to the designed or rated wind speed of a wind turbine and the closer their values the better. The most probable wind speed corresponds to the peak of the probability density function and represents the most likely wind speed to be experienced at a given site. The power produce by wind turbine can be estimated from [23]:

$$P(v) = \frac{1}{2} \rho A v_m^3 \tag{15}$$

where $\rho$ = the air density at the site (kg/m³), and A = the swept area of the rotor blades (m²). However, the wind power density (wind power per unit area) based on the Weibull probability density function can be calculated from [12] and [20]:

$$p(v) = \frac{P(v)}{A} = \frac{1}{2}\rho c^3 \Gamma\left(1 + \frac{3}{k}\right)$$

(16)

where p(v) is the wind power density (W/m³). In the case of Rayleigh distribution, the wind power density is given as [12] and [20]:

$$p(v) = \frac{3}{\pi}\rho v_m^3$$

(17)

## 5.2.2 SITE WIND CHARACTERISTICS

The site wind characteristics and Weibull parameters estimated using the above expressions for all the locations considered in the present study, are presented in Table 2. The table shows that the annual mean wind speeds are, respectively, 5.30 m/s, 4.50 m/s, 4.75 m/s, 4.51 m/s, 3.88 m/s and 4.00 m/s for Adafoah, Anloga, Aplaku, Mankoadze, Oshiyie and Warabeba.

**TABLE 2:** Annual and seasonal site wind speed characteristics and Weibull parameters at height 12 m height.

| Parameter | Adafoah | Anloga | Aplaku | Mankoadze | Oshiyie | Warabeba |
|---|---|---|---|---|---|---|
| v (m/s) | 5.30 | 4.50 | 4.75 | 4.51 | 3.88 | 4.00 |
| k | 2.00 | 2.00 | 2.47 | 2.21 | 2.27 | 1.92 |
| c (m/s) | 5.98 | 5.08 | 5.36 | 5.09 | 4.38 | 4.51 |
| vf (m/s) | 4.23 | 3.59 | 4.34 | 3.88 | 3.39 | 3.07 |
| ve (m/s) | 8.46 | 7.18 | 6.81 | 6.82 | 5.79 | 6.54 |
| p(v) (W/m²) | 174.86 | 107.19 | 105.19 | 98.04 | 61.10 | 78.46 |

Also shown in Table 2 is the value of the Weibull shape parameter for Aplaku, Mankoadze, Oshiyie and Warabeba previously determined directly and reported by Nkrumah [9]. It can be seen that the shape parameter varies between 1.92 and 2.47 for these sites. It is well known that the Rayleigh distribution can be used for the analysis for a site in absence of detailed wind speed data provided that mean wind speed data for desire resolution (hourly, daily, or monthly) is available. Based on this finding, the shape factor of 2 was assumed for Adafoah and Anloga. Similar assumption (k = 2) was previously used to roughly estimate the wind power potential in Ghana [9].

It can further be noted that the annual wind speed carrying maximum energy speeds are 8.46 m/s, 7.18 m/s, 6.81 m/s, 6.82 m/s, 5.79 m/s and 6.54 m/s for Adafoah, Anloga, Aplaku, Mankoadze, Oshiyie and Warabeba, respectively. For optimum performance, it was previously mentioned that the wind speed carrying maximum energy should be the same or close as possible to the rated wind speed. Thus, for all the locations, wind turbines with low to moderate designed or rated wind speed will be appropriate. As a result of Adafoah having highest value of wind speed carrying maximum energy, if wind turbines with the same design parameters are installed in all the sites considered in this study, the turbines will generate highest amount of electricity in Adafoah.

Furthermore, the annual mean power densities are, respectively, 174.86 W/m², 107.19 W/m², 105.19 W/m², 98.04 W/m², 61.10 W/m² and 78.46 W/m² for Adafoah, Anloga, Aplaku, Mankoadze, Oshiyie and Warabeba. Based on Pacific Northwest Laboratory (PNL) wind power classification scheme [24], the wind resource can roughly be classified into Class 2 (Adafoah, Anloga and Aplaku) and Class 1 (Mankoadze, Oshiyie and Warabeba) categories. In general, Class 1 wind resource category can marginally be considered for wind energy development while Class 1 can be considered unsuitable for large development but can be suitable for small scale applications. In addition, both Classes 1 and 2 can be used as part of hybrid energy system where wind power can be used in conjunction with other renewable energy resources (e.g., solar energy) and/or diesel/gasoline generators. It should be mentioned here that, the PNL classification scheme is based on measurement at height of 10 m. Detailed analysis and presentation of wind speed distribution in these locations and others can be found in [7] and [8].

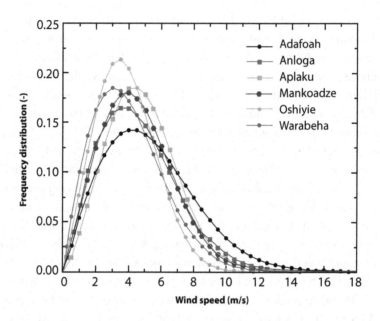

**FIGURE 2:** The annual Weibull probability density frequency for all the sites at height 12 m.

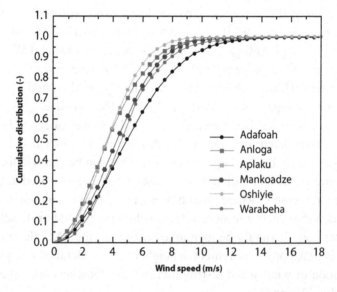

**FIGURE 3:** The annual Weibull cumulative frequency for all the sites at height 12 m.

## 5.2.3  WEIBULL FREQUENCY DISTRIBUTIONS

The annual Weibull probability density frequency and cumulative distributions of wind speed for the locations are respectively, shown in Fig. 2 and Fig. 3. The probability density function is used to illustrate the fraction of time for which a given wind speed can possibly prevail at a site. As indicated in Fig. 2, the most frequent wind speed expected in Adafoah, Anloga, Aplaku, Mankoadze, Oshiyie and Warabeba are about 4.0, 3.5, 4.5, 4.0, 3.5 and 3.5 m/s, respectively. It can be further observed that Adafoah has the highest spread of wind speed toward high wind speed while the Oshiyie has the lowest spread among these sites. Detailed comparison of the Weibull probability function with the actual wind data for some of the sites can be found in Nkrumah [9].

TABLE 3: Cumulative distribution as percentage (%) for selected wind speed at 12 m height.

| v (m/s) | F($\geq$ v) (%) | | | | | |
|---------|---------|--------|--------|-----------|---------|----------|
|         | Adafoah | Anloga | Aplaku | Mankoadze | Oshiyie | Warabeba |
| 2.00    | 89.4    | 86.1   | 91.6   | 88.1      | 84.5    | 81.1     |
| 2.50    | 84.0    | 79.2   | 85.9   | 81.2      | 75.6    | 72.5     |
| 3.00    | 77.8    | 71.4   | 78.8   | 73.3      | 65.5    | 63.3     |
| 3.50    | 71.0    | 63.3   | 70.5   | 64.6      | 54.8    | 54.1     |
| 4.00    | 63.9    | 55.0   | 61.5   | 55.6      | 44.3    | 45.2     |
| 4.50    | 56.8    | 46.9   | 52.2   | 46.7      | 34.5    | 36.9     |
| 5.00    | 49.7    | 39.3   | 43.1   | 38.2      | 25.9    | 29.6     |

The cumulative distribution function can be used for estimating the time for which wind speed is within a certain interval. For wind speeds (or cut-in wind speed) greater or equal to 2 m/s, Adafoah, Anloga, Aplaku, Mankoadze, Oshiyie and Warabeba are about 89.4, 86.1, 91.6, 88.1, 84.5 and 81.1 per cents, respectively, while the same locations, respectively, have frequencies of about 49.7, 39.3, 43.1, 38.2, 25.9 and 29.6 per cents for wind speed equal or greater than 5 m/s (see Fig. 3). Detailed frequencies for other possible cut-in wind speeds for all the sites considered are shown in Table 3. It can be concluded based on the information presented in this table that wind turbine conversion systems with designed cut-in wind speed of less than 3 m/s will be appropriate and better suited for electricity generation in these locations.

## 5.3 WIND TURBINE PERFORMANCE AND ECONOMIC ANALYSIS

### 5.3.1 POWER OUTPUT OF WIND TURBINE

A wind energy conversion system can operate at maximum efficiency only if it is designed for a particular site, because the rated power, cut-in and cut-off wind speeds must be defined based on the site wind characteristics [25]. It is essential that these parameters are selected so that energy output from the conversion system is maximized. The capacity factor $C_f$ is defined as the ratio of the mean power output to the rated electrical power ($P_{eR}$) of the wind turbine [21] and [25].

The performance of a wind turbine installed in a given site can be examined by the amount of mean power output over a period of time ($P_{out}$) and the conversion efficiency or capacity factor of the turbine. The mean power output Pout can be calculated using the following expression based on Weibull distribution function [25]:

$$P_{out} = P_{eR}\left\{\frac{e^{-\left(\frac{v_c}{c}\right)^k} - e^{-\left(\frac{v_r}{c}\right)^k}}{\left(\frac{v_r}{c}\right)^k - \left(\frac{v_c}{c}\right)^k} - e^{-\left(\frac{v_f}{c}\right)^k}\right\}$$

(18)

and the capacity factor $C_f$ of a wind turbine is given as:

$$C_f = \frac{P_{out}}{P_{eR}}$$

(19)

Therefore,

$$C_f = \left\{ \frac{e^{-\left(\frac{V_c}{c}\right)^k} - e^{-\left(\frac{V_r}{c}\right)^k}}{-\left(\frac{V_r}{c}\right)^k - \left(\frac{V_c}{c}\right)^k} - e^{-\left(\frac{V_f}{c}\right)^k} \right\}$$

(20)

where $v_c$, $v_r$, $v_f$ are the cut-in wind speed, rated wind speed and cut-off wind speed, respectively. It should be mentioned that the above expressions are only valid for the assessment of wind turbines that meet the following conditions: (1) the electrical power generated by the wind turbine is proportional to the wind speed in the region between the cut-in wind speed and the rated wind speed; and (2) the electrical power generated by the wind turbines is constant in the region between the rated wind speed and the cut-out wind speed. Eq. (20) indicates that the capacity factor is a function of the site parameters (which depend on the hub height) and the wind turbine design wind speed properties. The cost effectiveness of a wind turbine can be roughly estimated by the capacity factor of the turbine. This factor is a useful parameter for the both consumer and manufacturer of the wind turbine system [26].

In most cases, the available wind data are measured at height a different from the wind turbine hub height and since the wind speed at the hub height is of interest for wind power applications, the available wind speeds are adjusted to the wind turbine hub height using the following power law expression (e.g., [25] and [27]):

$$\frac{V}{V_0} = \left(\frac{h}{h_0}\right)^\alpha$$

(21)

where $V$ is the wind speed at the required height $h$, $V_0$ is wind speed at the original height $h_0$, and $\alpha$ is the surface roughness coefficient and is assumed to be 0.143 (or 1/7) in most cases. However, the surface roughness coefficient can also be determined from the following expression [26]:

$$\alpha = [0.37 - 0.088 \ln(V_0)]/ \left[1 - 0.088 \ln\left(\frac{h_0}{10}\right)\right]$$ (22)

Alternatively, the Weibull probability density function can be used to obtain the extrapolated values of wind speeds at different heights. This approach has been used in this study. Since the boundary layer development and the effect of the ground are non-linear with respect to the wind speed, the scale factor c and shape factor k of the Weibull distribution will change as a function of height by the following expressions [16]:

$$c(h) = c_0 \left(\frac{h}{h_0}\right)^n$$ (23)

$$k(h) = k_0 \left[1 - 0.088 \ln\left(\frac{h_0}{10}\right)\right] / \left[1 - 0.088 \ln\left(\frac{h}{10}\right)\right]$$ (24)

where $c_0$ and $k_0$ are the scale factor and shape parameter respectively at the measurement height h0 and h is the hub height. The exponent n is defined as:

$$n = [0.37 - 0.088 \ln(c_o)]/ \left[1 - 0.088 \ln\left(\frac{h}{10}\right)\right]$$ (25)

### 5.3.2 ECONOMIC ANALYSIS

The cost of electricity generated by a wind turbine depends on many factors which include: the site specific factors (e.g. wind speed and quantity of electricity generated, cost of land, and installation cost); cost of wind turbine and its economic life span; operation and maintenance; electricity tariff and incentives and exemptions [22] and [28]. Apart from the cost of the wind turbine which is set by the manufacturers, costs of other activities are location dependent. The cost of the wind turbine accounted for about 70% of the all total initial investment cost for wind energy development [22], [28] and [29]. As shown in Table 4, the specific cost of a wind turbine is dependent on the rated power and varies widely from one manufacturer to another [22], [28] and [30]. Since the economic feasibility of the wind energy development depends on its ability to generate electricity at a low operating cost per unit energy, accurate estimate of all the costs involved in generating electricity over the life span of the system is essential.

**TABLE 4:** Range of specific cost of wind turbines based on the rated power [31].

| Wind turbine size (kW) | Specific cost (kW) | Average specific cost (kW) |
| --- | --- | --- |
| <20 | 2200–3000 | 2600 |
| 20–200 | 1250–2300 | 1775 |
| >200 | 700–1600 | 1150 |

The commonly used method to estimate the operating cost of a unit energy produced by the wind energy conversion system is the levelised cost of electricity (LCOE) method [28]. The LCOE is a measure of the marginal cost of electricity over a period of time and it is commonly used to compare the electricity generation costs from various sources [29]. In addition, the LCOE gives the average electricity price needed for a net present value of zero when a discounted cash flow analysis is performed. The determination of the cost of unit energy involves three basic steps: (i) estimation of energy generated by the wind turbine over a given period (e.g. year); (ii) estimation of the total investment cost of the project; and (iii) division of the cost of investment by the energy produce by the system.

The unit cost of energy using the LCOE method can be estimated by the following expression [31]:

$$LCOE = \frac{CRF}{8760P_R C_f}\left(C_l + C_{om(esc)}\right) \text{ cost/kW h} \tag{26}$$

where $C_I$ is the total investment cost, $E_{WT}(=8760PRCf)$ is the annual energy output of wind turbine in kW h, CRF and $C_{om(esc)}$ are the capital recovery factor and present worth of the annual cost throughout lifetime of the wind turbines expressed as (10) and (11)[28] and [31]:

$$CRF = \frac{(1+\varepsilon)^n \varepsilon}{(1+\varepsilon)^n - 1} \tag{27}$$

$$C_{om(esc)} = \frac{C_{om}}{\varepsilon - e_{om}}\left[1 - \left(\frac{1+e_{om}}{1+\varepsilon}\right)^n\right] \text{cost/year} \tag{28}$$

where $C_{om}$, $e_{om}$, n and ε are the operation and maintenance cost for the first year, escalation of operation and maintenance, useful lifetime of turbine and discount rate respectively. The discount rate can be corrected for inflation rate (r) and inflation escalation rate (e) using the following expressions [22]:

$$e_a = \{(1+e)(1+r)\} - 1 \tag{29}$$

where $e_a$ is called the apparent escalation rate. The discount rate can be determined from:

$$\varepsilon = \frac{(1+i)}{(1+e_a)} - 1 \tag{30}$$

## 5.4 RESULTS AND DISCUSSION

For wind turbine performance assessment and economic analysis, four small commercial wind turbine models: Polaris 15–50 [32], CF-100 [33],

Garbi-150/28 [34] and WES 30 [35] with rated power range from 50 kW to 250 kW are selected. These wind turbines were selected due to their relatively low cut-in wind speeds (range between 2.2 m/s and 2.7 m/s) and moderate rated wind speeds (range between 9.5 m/s and 13 m/s). The chosen wind turbines are appropriate for site with low to medium wind speed regimes, such as those selected for this study. In addition, these wind turbines satisfied the conditions listed in Section 3.1 and therefore, Eqs. (18), (19) and (20) were used for estimating the power output and the capacity factor for each turbine for a given site. The selected wind turbine models and their design properties are presented in Table 5 and their power curves are shown in Fig. 4 and Fig. 5. It should be mentioned here that these wind turbines are designed to operate at different hub heights. However, the highest hub heights for each model were used, since the wind speed increases with the wind turbine hub height. For each location, the annual energy output by each turbine and capacity factor based on Weibull distribution function parameters were determined and presented in the following sections.

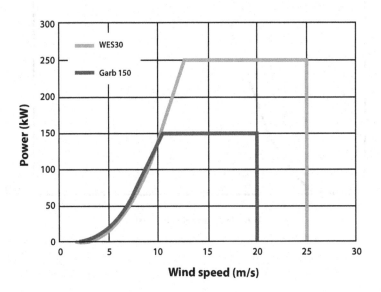

**FIGURE 4:** The power-speed curves for WES30 and Garbi-150 wind turbine models.

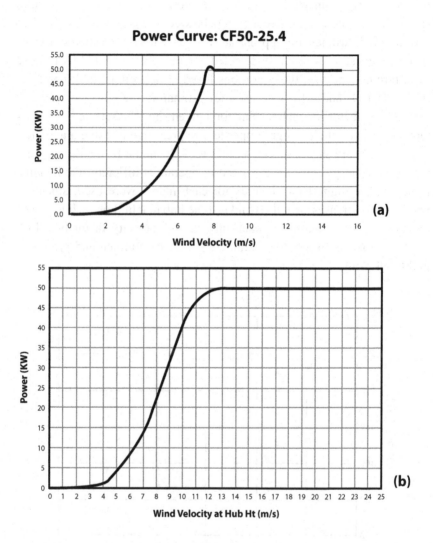

**FIGURE 5:** The power-speed curves for (a) CF-100 wind turbine model [33] and (b) Polaris 15–50 wind turbine model [32].

**TABLE 5:** Specifications of the selected wind turbine models [33], [34], [35] and [36].

|  | WES30 | CF-100 | Polaris 15–50 | Garbi150/28 |
|---|---|---|---|---|
| Rated Power (kW) | 250.0 | 100.0 | 50.0 | 150.0 |
| Hub height (m) | 48.0 | 34.4 | 50 | 30.3 |
| Rotor diameter (m) | 30.0 | 25.4 | 15.2 | 28.0 |
| Cut-in wind speed (m/s) | 2.3 | 2.2 | 2.7 | 2.5 |
| Rated wind speed (m/s) | 13.0 | 9.5 | 10 | 10.4 |
| Cut-off wind speed (m/s) | 25.0 | 35.0 | 25.0 | 20.0 |

## 5.4.1 WIND TURBINE PERFORMANCE

### 5.4.1.1 ANNUAL ELECTRICITY OUTPUT

The annual energy output from the selected wind turbine models at all the locations is presented in Fig. 6. The annual energy output ranges from about 128 MWh in Oshiyie using the Polaris15–50 model to 730 MWh in Adafoah using the WES30 model. Irrespective of the location, the WES30 wind turbine model generated highest amount of electricity while the Polaris15–50 model generated least electricity. This is expected due to the large size of the WES30 model when compared with other models. Based only on the amount of electricity produced, the WES30 model may be the best choice for all the locations considered in this study. However, as will be shown later, the decision to select a particular wind turbine should not only be based on its size, but also on its ability to produce electricity at the cheapest cost.

In addition to the total energy output, the capacity factor of a wind turbine can be used to access economic viability of a wind energy conversion systems project. For an investment in wind power to be cost effective, a system with a capacity factor greater than 0.25 is generally recommended [19] and [22]. The capacity factors for the selected wind turbine models are presented in Fig. 7. The Polaris15–50 model has the highest value among the models considered for all the sites. This is partly due to its low cut-in wind speed of 2.7 m/s and most importantly, its rated wind speed of 10 m/s. The Cf values for this model are 50.6, 41.8, 41.0, 39.4, 29.2 and 34.3

per cents for Adafoah, Anloga, Aplaku, Mankoadze, Oshiyie and Wara-beba, respectively. Due to its low cut-in and rated wind speed, the capacity factors of CF-100 are close to those of Polaris 15–50 model even though its hub height is about 69% of that Polaris 15–50 model. It can further be observed that despite the fact that the WES30 model have comparable hub height with Polaris 15–50 (see Table 5), it has the least capacity factors for each site. This is primarily due to its relatively high rated wind speed (of 13 m/s) compared with other wind turbine models. Hence, for wind tur-bines with comparable hub heights, the rated wind speed is a more impor-tant parameter than the hub height. In general, estimated capacity factors indicate that an investment in wind energy development can be considered worthwhile in all the sites using the CF-100 and Polaris 15–50 models or wind turbine with similar designed parameters. Other wind turbine models have high capacity factors in most of the sites and therefore, can also be considered appropriate for wind power development in those sites.

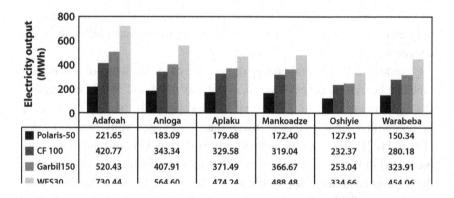

**FIGURE 6:** Estimated electricity output from selected wind turbine models.

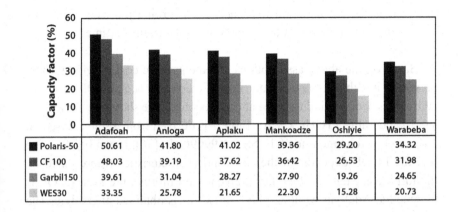

| | Adafoah | Anloga | Aplaku | Mankoadze | Oshiyie | Warabeba |
|---|---|---|---|---|---|---|
| ■ Polaris-50 | 50.61 | 41.80 | 41.02 | 39.36 | 29.20 | 34.32 |
| ■ CF 100 | 48.03 | 39.19 | 37.62 | 36.42 | 26.53 | 31.98 |
| ■ Garbil150 | 39.61 | 31.04 | 28.27 | 27.90 | 19.26 | 24.65 |
| WES30 | 33.35 | 25.78 | 21.65 | 22.30 | 15.28 | 20.73 |

**FIGURE 7:** The capacity factors for the selected wind turbine models.

## 5.4.2 COST OF ELECTRICITY

In estimating the levelised cost of electricity generated by each selected commercial wind turbine at the selected locations, the following assumptions were taken into consideration:

1. The lifetime (n) of each wind turbine is taken to be 20 years.
2. According to the available information (Bank of Ghana [36]), the current interest rate (r) and inflation rate (i) in Ghana are 15% and 10.4% respectively and these values were used in this analysis.
3. The actual total of other initial costs depends, among other factors, on the available infrastructure in the selected locations. In developing countries, the cost of infrastructure, installation, and grid connection represents about 30–50% of the total capital cost [27]. In this study, therefore, other initial civil work costs are taken to be 40% of the total initial cost [37].
4. The annual operation and maintenance costs are assumed to be 7% of the initial capital cost of the wind turbine installation system (system price/lifetime). This value is consistent with the range of 1–7% reported by Manwell et al. [27].

5. It is further assumed that each wind turbine produces the same amount of energy in each year during its useful lifetime.

Based on the above assumptions and the specific cost of wind turbine presented in Table 4, the estimated cost of electricity (i.e., LCOE) generated by the selected wind turbines at each locations are shown in Table 6 ($/kW h) and Table 7 (GH¢/kW h). Table 7 is derived from Table 6 based on the current exchange rate of 1US$ = 1.9909GH¢ [36]. It can be observed that the LCOE depends on the site wind characteristics (represented by the turbine capacity factor). For a given wind turbine, the least LCOE is obtained at Adafoah while the highest LCOE is obtained at Oshiyie. The least cost of unit energy per kW h is obtained with the Polaris 15–50 model as 3.50 $cents/kW h, 4.24 $cents/kW h, 4.32 $cents/kW h, 4.51 $cents/kW h, 6.07 $cents/kW h and 5.17 $cents/kW h for Adafoah, Anloga, Aplaku, Mankoadze and Warabeba. It can further be observed that the highest cost of electricity is obtained by the Garbi-150 model with cost ranging between 6.91 $cents/kW h at Adafoah and 11.10 $cents/kW h at Oshiyie.

**TABLE 6:** Cost analysis for selected wind turbine models ($/kW h).

| Location | Adafoah | Anloga | Aplaku | Mankoadze | Oshiyie | Warabeba |
|----------|---------|--------|--------|-----------|---------|----------|
| Polaris-50 | 0.0350 | 0.0424 | 0.0432 | 0.0451 | 0.0607 | 0.0517 |
| CF 100 | 0.0369 | 0.0452 | 0.0471 | 0.0487 | 0.0668 | 0.0554 |
| Garbi150 | 0.0691 | 0.0882 | 0.0968 | 0.0981 | 0.1421 | 0.1110 |
| WES30 | 0.0532 | 0.0688 | 0.0819 | 0.0795 | 0.1160 | 0.0855 |

**TABLE 7:** Cost analysis for selected wind turbine models (GH¢/kW h).

| Location | Adafoah | Anloga | Aplaku | Mankoadze | Oshiyie | Warabeba |
|----------|---------|--------|--------|-----------|---------|----------|
| Polaris-50 | 0.0695 | 0.0841 | 0.0857 | 0.0893 | 0.1204 | 0.1024 |
| CF 100 | 0.0732 | 0.0897 | 0.0934 | 0.0965 | 0.1325 | 0.1099 |
| Garbi150 | 0.1370 | 0.1748 | 0.1919 | 0.1944 | 0.2817 | 0.2201 |
| WES30 | 0.1054 | 0.1363 | 0.1623 | 0.1576 | 0.2300 | 0.1695 |

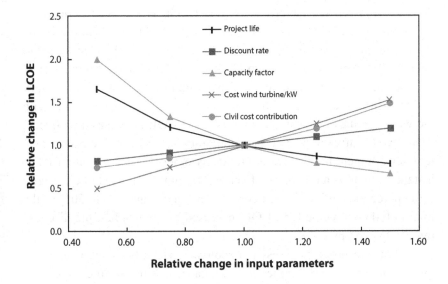

**FIGURE 8:** Sensitivity analysis of selected input parameters of LCOE for Polaris 15–50 wind turbine model in Oshiyie.

In order to investigate the effects of uncertainty or changes in some of the input parameters (such as the unit cost of wind turbine, discount rate, useful (project) life of the wind turbine, other initial capital cost and wind turbine capacity factor (which is a function of the site wind characteristics) on the unit cost of electricity, sensitivity analysis was also carried out on Polaris 15–50 wind turbine model in Oshiyie (as a case study). The results of this analysis are shown in Fig. 8. It can be noted that the degree of influence of these on parameters on LCOE is not the same. It can be observed from this figure that the LCOE is sensitive to all the input parameters. The degree of sensitivity can be classified into two different groups. The first group comprises the capacity factor and useful life of the wind turbine. These parameters affect the LCOE very positively, that is, LCOE decreases as their values increase. This explains why a site with wind resource that provides high wind turbine capacity factor is desirable (economically). When the capacity factor is improved by 25% (for instance, due to better matching of the selected wind turbine with the site wind speed characteristics), the LCOE is observed to decrease by 20% (from US$0.06072/kW h to US$0.04858/kW h). Similarly, by increasing the project life from 20 to 25 years, the LCOE is observed to decrease by about 12.47% (from US$0.06072/kW h to US$0.05315/kW h). The second group is made up of the wind turbine cost, the civil work and infrastructure costs, the discount rate. This figure indicates that with increasing values of these parameters, the LCOE increases, and hence, they have negative impact on the economic viability of wind energy system development in this location (and any other location in general). For instances, there is an increase of about 20% in the unit cost of electricity if the percentage of civil work cost is increased from 40% to 50% of the total initial cost, while the LCOE increased by about 1% if the discount rate is changed by 25%.

In order to properly assess the economic viability of the wind energy conversion systems application in Ghana, it is useful to compare the cost of electricity generated using the selected wind turbines with current electricity tariffs in Ghana. The electricity tariffs in Ghana is set and regulated by the Public Utilities Regulatory Commission (PURC). The average electricity end user tariffs for 2012 was GH¢ 0.232/kW h [4]. It should be mentioned that the government contends that these prices are heavily subsidized. As a result the PURC has requested an upward adjustment in the bulk generation tariff to reflect increases in costs of power supply for 2013 [38]. Therefore, comparing this value with the estimated cost presented in Table 7, the following observation can be made: (1) the wind turbine model CF-100 can be considered to be the most economically viable options at all selected sites; (2) all the wind turbines can be considered viable at Adafoah, and (3) other wind turbine models can be considered as marginally economically viable at the other locations. With current governments' desire to increase the contribution of renewable energy in electricity generation, appropriate financial assistance in terms of subsidy and development of infrastructure by government, can effectively reduce the total initial investment cost and encourage investment in wind energy development in Ghana.

## 5.5 CONCLUSIONS

In this study, the techno-economic analysis of wind energy development in six selected sites along the coastal region of Ghana was examined. The findings from this study can be summarized as follows:

1. The wind resource along the coastal region of Ghana can be classified into Class 2 or less wind resource category. This makes wind resource in this area marginally suitable for large scale wind energy development or suitable for small scale applications and can be useful part of hybrid energy system.
2. The wind turbine model CF-100 (with cut-in wind of 2.2 m/s and rated wind speed of 9.5 m/s) gave the unit cost of electricity range between 3.69 $cents/kW h and 6.68 $cents/kW h and is considered the most economically viable options for all the sites.
3. Wind turbine with designed cut-in wind speed of less than 3 m/s and moderate rated wind speed between 9 and 11 m/s will be suitable for wind energy development along the coastal region of Ghana.
4. With government assistance in the form of subsidy and development of infrastructure investment in wind energy development can more economically viable. In addition, government assistance in the area of manpower training for skills knowledge on wind energy technology (from design to production stage using locally available resources) will accelerate the wind energy development in Ghana.

## REFERENCES

1. Global Wind Energy Council. Global wind energy report: Annual market update; 2012.
2. B. BoroumandJazi, B. Rismanchi, R. Saidur. Technical characteristic analysis of wind energy conversion systems for sustainable development. Energy Convers Manage, 69 (2013), pp. 87–94
3. G.L. Park, B.S. Richards, A.I. Schäfer. Potential of wind-powered renewable energy membrane systems for Ghana. Desalination, 248 (2009), pp. 169–176
4. National Energy Statistics. 2000–2012. Energy Commission, Ghana; 2013.

5.  Ministry of Energy and Petroleum, Ghana. Achievement of Power Directorate; 2013. <http://www.energymin.gov.gh/?page_id=183> (accessed 20.06.13).

6.  Volta River Authority, Ghana. VRA completes first solar plant; 2013. <http://www.vra.com>.

7.  Energy commission, Ghana. Solar and wind energy resource assessment (SWERA); 2003. Available on: <http://www.energycenter.knust.edu.gh/downloads/5/5598.pdf>.

8.  Nkrumah F. Feasibility study of wind energy utilization along the coast of Ghana. MSc Thesis. KNUST, Ghana; 2002. Available on: <http://www.energycenter.knust.edu.gh/downloads/5/5782.pdf>.

9.  Ghana Wind Energy Resource Mapping Activity. National Renewable Energy Laboratory. Technical Report. USDOE; 2003. Available on: <http://www.en.openei.org/datasets/files/717/pub/ghanawindreport_245.pdf>.

10. S.O. Oyedepo, M.S. Adaramola, S.S. Paul. Analysis of wind speed data and wind energy potential in three selected locations in South East Nigeria. Int J Energy Environ Eng, 3 (2012), p. 7

11. S.A. Akdag, H.S. Bagiorgas, G. Mihalakakou. Use of two-component Weibull mixtures in the analysis of wind speed in the eastern Mediterranean. Appl Energy, 87 (2010), pp. 2566–2573

12. O.S. Ohunakin, M.S. Adaramola, O.M. Oyewola. Wind energy evaluation for electricity generation using WECS in seven selected locations in Nigeria. Appl Energy, 88 (2011), pp. 3197–3206

13. M. Mpholo, T. Mathaba, M. Letuma. Wind profile assessment at Masitise and Sani in Lesotho for potential off-grid electricity generation. Energy Convers Manage, 53 (2012), pp. 118–127

14. A. Mostfaeipour. Economic evaluation of small wind turbine utilization in Kerman, Iran. Energy Convers Manage, 73 (2013), pp. 214–225

15. I. Fyrippis, P.J. Axaopoulos, G. Panayiotou. Wind energy potential assessment in Naxos Island, Greece. Appl Energy, 87 (2010), pp. 577–586

16. C.G. Justus, W.R. Hargraves, A. Mikhail, D. Graber. Methods for estimating wind speed frequency distributions. J Appl Meteorol, 17 (1978), pp. 350–353

17. S.D. Kwon. Uncertainty analysis of wind energy potential assessment. Appl Energy, 87 (2010), pp. 856–865

18. F.A.L. Jowder. Wind power analysis and site matching of wind turbine generators in Kingdom of Bahrain. Appl Energy, 86 (2009), pp. 538–545

19. M.S. Adaramola, O.M. Oyewola. Evaluating the performance of wind turbines in selected locations in Oyo state, Nigeria. Renew Energy, 36 (2011), pp. 3297–3304

20. T. Maatallah, S. El-Alimi, A.W. Dahmouni, A.B. Nasrallah. Wind power assessment and evaluation of electricity generation in the Gulf of Tunis, Tunisia. Sustain Cities Society, 6 (2013), pp. 1–10

21. A. Balouktsis, D. Chassapis, T.D. Karapantsios. A nomogram method for estimating the energy produced by wind turbine generators. Solar Energy, 72 (2002), pp. 251–259

22. S. Mathew. Wind energy: fundamentals, resource analysis and economics. Springer, Heidelberg (2006)

23. A.S. Ahmed. Potential wind power generation in South Egypt. Renew Sustain Energy Rev, 16 (2012), pp. 1528–1536

24. A. Ilinca, E. McCarthy, J.-L. Chaumel, J.-L. Retiveau. Wind potential assessment of Quebec Province. Renew Energy, 28 (2003), pp. 1881–1897

25. E.K. Akpinar, S. Akpinar. An assessment on seasonal analysis of wind energy characteristics and wind turbine characteristics. Energy Convers Manage, 46 (2005), pp. 1848–1867

26. A. Ucar, F. Balo. Evaluation of wind energy potential and electricity generation at six locations in Turkey. Appl Energy, 86 (2009), pp. 1864–1872

27. J.F. Manwell, J.G. McGowan, A.L. Rogers. Wind energy explained: theory, design and application. (2nd ed.)John Wiley and Sons Ltd, Chichester (2009)

28. M. Gökçek, M.S. Genç. Evaluation of electricity generation and energy cost of wind energy conversion systems (WECSs) in Central Turkey. Appl Energy, 86 (2009), pp. 2731–2739

29. Hearps P, McConell D. Renewable energy technology cost review. Technical Paper Series, Melbourne Energy Institute; March 2011.

30. M. Gökçek, H.H. Erdem, A. Bayülken. A techno-economical evaluation for installation of suitable wind energy plants in western Marmara, Turkey. Energy Exploration Exploitation, 25 (2007), pp. 407–428

31. M.S. Adaramola, S.S. Paul, S.O. Oyedepo. Assessment of electricity generation and energy cost of wind energy conversion systems in north-central Nigeria. Energy Convers Manage, 52 (2011), pp. 3363–3368

32. <http://www.polarisamerica.com/turbines/>.

33. <http://www.mint-energy.co.uk/50-75-100>.

34. <http://www.ardnell.com/>.

35. <http://www.windenergysolutions.nl/wes-250>.

36. Bank of Ghana: <http://www.bog.gov.gh/> (accessed 13.05.13).

37. O.S. Ohunakin, O.M. Oyewola, M.S. Adaramola. Economic analysis of wind energy conversion systems using levelised cost of electricity and present value cost methods in Nigeria. Int J Energy Environ Eng, 4 (2013), p. 2

38. Public Utilities Regulatory Commission, Ghana. Template for filing of tariff proposals by generation companies; 2013.

# CHAPTER 6

# Assessing Environmental Impacts of Offshore Wind Farms: Lessons Learned and Recommendations for the Future

## HELEN BAILEY, KATE L. BROOKES, AND PAUL M. THOMPSON

## 6.1 INTRODUCTION

Efforts to reduce carbon emissions and increase production from renewable energy sources have led to rapid growth in offshore wind energy generation, particularly in northern European waters [1,2]. The first commercial scale offshore wind farm, Horns Rev 1 (160 MW with 80 turbines of 2 MW), became operational in 2002. The average capacity of turbines and size of offshore wind farms have been increasing since then, and they are being installed in deeper waters further from the coast. By the end of 2013, operational wind farms were in an average water depth of 16 m and 29 km from shore in Europe [3] (Figure 1). With technological advances in the future [4] there is likely to be a continued increase in the size of offshore wind projects [3], but there are still uncertainties about the effects on the environment [5]. The novelty of the technology and construction processes make it difficult to identify all of the stressors on marine species and to estimate the effect of these activities [6].

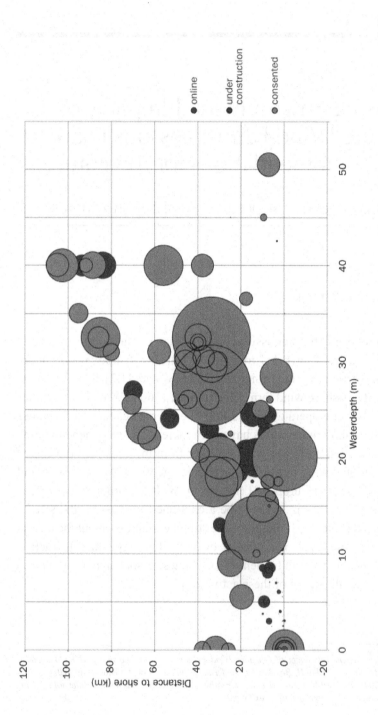

**FIGURE 1:** Average water depth and distance to shore of offshore wind farms (reproduced from ref. 3, source EWEA). Operational (online), under construction and consented wind farms in Europe up to the end of 2013 are occurring at increasing water depths and distances from shore. The circle size represents the total power capacity of the wind farm.

The major environmental concerns related to offshore wind developments are increased noise levels, risk of collisions, changes to benthic and pelagic habitats, alterations to food webs, and pollution from increased vessel traffic or release of contaminants from seabed sediments. There are several reviews of the potential impacts of offshore wind energy on marine species e.g. [5-7]. As well as potential adverse impacts, there are possible environmental benefits. For example, wind turbine foundations may act as artificial reefs, providing a surface to which animals attach. Consequently there can be increases in the number of shellfish, and the animals that feed on them, including fish and marine mammals [8-11]. A second possible benefit is the sheltering effect. A safety buffer zone surrounding the wind turbines may become a de-facto marine reserve, as the exclusion of boats within this zone would reduce disturbance from shipping. Exclusion of some or all types of fishing could also result in local increases in prey abundance for top predators, whilst reducing the risk of bycatch in fishing gear [9]. Further research is required to understand the ability of wind turbines to attract marine species and the effect of excluding fisheries. Finally, there may also be opportunities in the future to combine offshore wind farms with open ocean aquaculture [12].

Over 2,000 wind turbines are installed in 69 offshore wind farms across Europe, with the greatest installed capacity currently in the U.K. (Figure 2) [3]. As the number of offshore wind farms has increased, approaches for environmental monitoring and assessment have improved over time. However, there are still few studies that have measured the responses of marine species to offshore wind farm construction and operation, and none have yet assessed longer term impacts at the population level. In Europe, legislation requires consideration of cumulative impacts, defined as impacts that result from incremental changes caused by other past, present or foreseeable actions together with the project [13]. However, approaches for cumulative impact assessments currently vary in terms of their transparency, efficiency and complexity, and this is an active area of research development [14]. In addition to assessing and measuring impacts, it is also necessary to develop decision support tools that will assist regulators with determining whether a proposed development can be legally consented.

**FIGURE 2:** Offshore wind farms around the U.K., July 2014. This includes wind farms in operation (black wind turbines), those consented and under development (blue stars), and in the proposal and planning stages (red stars).

In this paper, we first briefly review the potential impacts of offshore wind developments on marine species. We then identify the key lessons that have been learned from our own studies and others in Europe, primarily focusing on marine mammals and seabirds. Much of the environmental research that has been conducted in relation to offshore wind energy has concerned the impact of sound exposure for marine mammals and the risk of collisions with turbines for seabirds. We identify where knowledge gaps exist that could help to improve current models and impact assessments. Finally, we discuss emerging technologies and make recommendations for future research to support regulators, developers and researchers involved in proposed developments, particularly in countries where the implementation of offshore wind energy is still in its early stages.

## 6.1.1 IMPACT PATHWAYS

The potential effects of offshore wind farm construction and operation will differ among species, depending on their likelihood of interaction with the structures and cables, sensitivities, and avoidance responses. Studies have generally focused on marine mammals and seabirds because of stakeholder concerns and legal protection for these species and their habitats. The construction phase is likely to have the greatest impact on marine mammals and the activities of greatest concern are pile driving and increased vessel traffic [15]. Pile driving is currently the most common method used to secure the turbine foundation to the seafloor, although other foundation types are being developed [4]. The loud sounds emitted during pile driving could potentially cause hearing damage, masking of calls or spatial displacement as animals move out of the area to avoid the noise [16,17]. Fish could similarly be affected by these sounds [17-20]. There is also a risk to marine mammals, sea turtles and fish of collision and disturbance from vessel movements associated with surveying and installation activities.

During operation of the wind turbines, underwater sound levels are unlikely to reach dangerous levels or mask acoustic communication of marine mammals [21,22]. However, this phase of the development is of greatest concern for seabirds. Mortality can be caused by collision with the moving turbine blades, and avoidance responses may result in displacement from key habitat or increase energetic costs [23,24]. This may affect birds migrating through the area as well as those that breed or forage in the vicinity.

During operation, cables transmitting the produced electricity will also emit electromagnetic fields. This could affect the movements and navigation of species that are sensitive to electro- or magnetic fields, which includes fish species, particularly elasmobranchs and some teleost fish and decapod crustaceans, and sea turtles [25-27]. Commercial fish species may potentially be positively affected if fishing is prohibited in the vicinity of the wind farm, although this could result in a displacement of fisheries effort and consequent change in catches and bycatch.

The specific species of greatest concern will differ among regions depending on their occurrence and protection status. For example, assessments of impacts upon marine mammals in Europe have generally focused on small cetaceans (particularly harbor porpoises (*Phocoena phocoena*)) and pinnipeds (primarily harbor seals (*Phoca vitulina*)). These species are common in such areas and the EU Habitats Directive (92/43/EEC) requires governments to establish Special Areas of Conservation for their protection. However, in other locations, marine mammal species listed under the Endangered Species Act, such as the North Atlantic right whale (*Eubalaena glacialis*), blue whale (*Balaenoptera musculus*), humpback whale (*Megaptera novaeangliae*), and fin whale (*Balaenoptera physalus*), may be of greater concern. Based on their call frequencies, these large whales are considered to be sensitive to the low frequency sounds produced during pile driving [16,28,29].

There is also a paucity of information on the effects of human-generated sound on fish [18,20,30,31]. Evidence of injury from pile driving sounds in a laboratory simulated environment has been reported for several fish species [32-34]. Recovery tended to occur within 10 days of exposure and is unlikely to have affected the survival of the exposed animals. Common sole larvae (*Solea solea*) also survived high levels of pile-driving sound in controlled exposure experiments [35]. However, a behavioral response was triggered in cod (*Gadus morhua*) and sole by playbacks of pile driving sounds in the field and was initiated at a much lower received sound level [36]. This could consequently result in a large zone of behavioral response. The sounds produced by offshore wind farms may also mask fish communication and orientation signals [30]. These responses need to be investigated further to determine their potential effect on foraging, breeding and migration, and require the ability to record the movements of fish as well as the measurement of sound pressure levels and particle motion since fish are sensitive to both [18,37]. Some fish species are short-lived and highly fecund reducing the likelihood of any longer-term population level effects from wind farm noise and disturbance. However, this is not true of all fish species and there are endangered species, such as the Atlantic sturgeon (*Acipenser oxyrinchus*), those listed as vulnerable, such as the basking shark (*Cetorhinus maximus*), and potential impacts to fisheries which may need to be taken into consideration.

Much of the early work investigating impacts upon bird populations at European sites has focused on species of migratory or wintering waterfowl [23,38]. There is much less known about potential collision risk or displacement for the broader suite of seabird species that occur in many of the areas currently being considered for large scale wind farm developments. Migrating bats have also been found to occur offshore [39,40], although relatively little research on their offshore distribution, collision risk and potential displacement by offshore wind farms has been done compared to that for wind farms on land [39,41].

Other taxonomic groups such as sea turtles are rare visitors to coastal European waters, and have not been considered at high risk from the effects of offshore wind farms. However, in other areas, for example along the North American coast, there may be sea turtle nesting or breeding grounds in the vicinity of proposed sites [42]. It has recently been determined that the hearing sensitivity of leatherback turtles overlaps with the frequencies and source levels produced by many anthropogenic sounds, including pile-driving [43]. This highlights the need for a better understanding of the potential physiological and behavioral impacts on sea turtles.

### 6.1.2 LESSONS LEARNED

Environmental research for offshore wind energy has evolved over time in Europe as a better understanding of the type of information and analysis that best informs decisions about the siting of offshore wind facilities has been developed. Other countries interested in offshore wind energy, such as the U.S.A. (Figure 3), may therefore benefit from the European experience and hindsight to maximize the potential success of their projects [44]. Based on our experiences relating to marine mammals and seabirds, the key lessons learned that we have identified are:

1.  Define the area of potential effect

- Identifying the area over which biological effects may occur to inform baseline data collection.
- Determining the connectivity between key populations and proposed wind energy sites.

2.  Identify the scale and significance of population level impacts

- The need to define populations, identify which populations occur within the wind energy site and the area of potential effect, and their current status.
- The requirement for demographic data and information on vital rates to link individual responses to population level consequences.
- Research to test and validate modeling assumptions and parameters.

3.  Validate models through measuring responses in the field

- Use of a gradient design to determine the extent of spatial displacement as a result of offshore wind energy development and how this may change over time.
- Utilization of techniques with the power to detect changes.
- Coordination of human activities and monitoring in the vicinity of wind energy sites.

4.  Learn from other industries to inform risk assessments and the effectiveness of mitigation measures.

- Onshore wind energy, seismic surveys and floating oil platforms.

We will discuss each of these lessons learned in more detail and then provide recommendations for further research to fill identified knowledge gaps, test existing models and improve future environmental impact assessments (EIA).

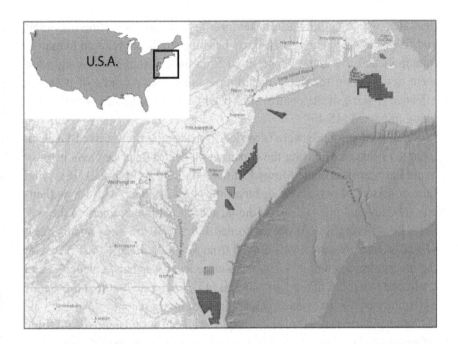

**FIGURE 3:** Potential wind energy areas in the Mid-Atlantic off the U.S.A.(source BOEM). There are no commercial wind farms currently on the Atlantic outer continental shelf off the U.S.A., but the Bureau of Ocean Energy Management (BOEM) has designated wind energy areas, which are in the planning (orange squares) and leased (green squares) stages (July 2014).

## 6.1.3 THE AREA OF POTENTIAL EFFECT

To evaluate the impact of a proposed activity on marine species, it is necessary to have sufficient baseline information on distribution, abundance and their trends within the area of potential effect. This is particularly challenging for many marine species since some stressors, such as underwater sound, can travel long distances, and these species are often highly mobile and/or migratory. Consequently, the area of potential effect can extend far beyond the immediate vicinity of the proposed development. For example, the sound produced during pile driving may travel tens of kilometers underwater, which could cause behavioral disturbance to marine mammals and fish over a large area [17]. Early baseline studies in relation to marine mammals and the impact upon them from underwater pile driving noise were designed on relatively uncertain estimates of the area of potential effect, and some of the control sites used were subsequently identified as being impacted [45,46].

Pile-driving sounds were recorded to determine the received levels at a range of distances during the construction of the Beatrice Demonstrator Wind Farm project in the Moray Firth, Scotland [47] (Figure 2). This wind farm consisted of two 5 MW turbines in water 42 m deep with 1.8 m diameter tubular steel piles to secure the jacket foundation structure to the seafloor. Pile-driving sounds were recorded at 0.1 to 70 km away and the peak to peak sound pressure levels ranged from 122–205 dB re 1 µPa. Based on the measurements within 1 km, the source level was back-calculated and estimated to be 226 dB re 1 µPa at 1 m [47]. Each pile required thousands of hammer blows and was struck about once every second.

Since pile-driving involves multiple strikes, it is considered a multiple pulse sound. The sound exposure level (SEL) is a measure of the energy of a sound and depends on both the pressure level and duration [28]. This can be summed over multiple strikes to give the cumulative SEL [19,28]. The cumulative energy level over the full pile-driving duration gives a measure of the dose of exposure, assuming no recovery of hearing between repeated strikes, and is necessary for assessing cumulative impacts.

There were, and still are, insufficient data available to develop noise exposure criteria for behavioral responses of marine mammals to multiple-pulsed noise such as pile driving [28]. Evidence of behavioral disturbance from sounds arising from pile-driving has been obtained through simulated, playback, and live conditions, and indicates that the zone of responsiveness for harbor porpoises may extend to 20 km or more [45,46,48-52]. However, response distances will vary depending on the activity being undertaken by the animal when it is exposed to the sound, the sound source level, sound propagation, and ambient noise levels [16,28,53]. Limited understanding of the role of these different environmental, physical and biological factors currently constrain assessments of the potential scale of impact at particular development sites.

Collecting baseline information for such large areas of potential effect presents a number of challenges. In cases where there is little or no existing information on the species of concern, such as their distribution and abundance in more offshore areas, it can be difficult to determine appropriate designs for impact studies [49]. The logistical difficulties of working offshore, along with financial limitations, may additionally restrict the number of sampling sites and replicates. It has been recommended that at least two years of baseline data are necessary for a sufficient description of species occurrences [54]. However, whilst this may provide information on seasonal variability, longer time-series of data are ideally required to capture inter-annual variability in order to identify the effects of construction activities over natural variation (which may be high) [54,55]. Given that data collection over these large spatial and temporal scales will be so difficult to achieve, it is crucial that studies are targeted to focus upon those data which are critical for supporting decision making. Baseline data and modeling should be targeted to answer specific questions relating to the consenting process, meaning that monitoring and site characterization requirements may differ under different legislative systems. For example, consent may require an understanding of the connectivity of a proposed wind energy site with key protected populations. In Scottish waters, this requirement has focused research efforts on areas within and surrounding proposed wind energy sites, and between these sites and EU designated Special Areas of Conservation (SACs) for harbor seals and for bottlenose dolphins [56-58].

For birds, the operational phase of wind farms is likely to present the biggest risk. Vulnerability and mortality at onshore wind turbines has been identified as being related to a combination of site-specific, species-specific and seasonal factors [59]. The development of collision risk models for seabirds requires information on their spatial distribution and flight heights to determine the likelihood of co-occurrence with the wind turbine blades, and their avoidance response to estimate the mortality risk [60]. However, much of this relies on expert-based estimates because there are very few empirical data on flight heights for different seabird species [24]. Although there have been estimates of flight heights during ship-based surveys where they are classified into altitude categories there can be large inter-observer differences [61]. One recent approach to address this data gap is to model flight height distributions based on compilations of survey data [62,63].

While information on site-specific flight heights of bird species is lacking, there is even less information on avoidance responses to large offshore wind farms by birds. The few studies examining avoidance behavior involved tracking eider ducks (*Somateria mollissima*) and geese by radar. These studies documented a substantial avoidance response by these migrating birds, which reduced the collision risk [23,64]. There is a need for empirical data on both broad and fine-scale avoidance responses to improve the reliability of predictions from collision risk models [65]. There should also be a focus not only on estimates of mortality, but also of the energetic consequences of avoidance and displacement behaviors [66], and their impacts on survival and fecundity. The cumulative impacts of different disturbance activities (such as ship and helicopter traffic) and multiple wind energy sites within the migration pathway or home range of a population should also be considered [67].

### 6.1.4 POPULATION LEVEL IMPACTS

The regulatory requirements for assessing the impacts of a proposed activity and determining whether it is biologically significant will vary among countries. However, in general this process will require populations to be defined, identifying which of these populations occur within the area of potential effect, and understanding their current status to determine whether the impact will be significant. The complexity of approaches, models and simulation tools to support these assessments has greatly increased over time [24,58,68]. However, there are still many knowledge gaps concerning behavioral responses, particularly on the consequences of any behavioral change on vital rates. For example, there is a growing understanding that anthropogenic noise, such as pile-driving, may affect the behavior of marine mammals and lead to spatial displacement [69]. However, there have been no empirical studies linking the consequences of this behavioral response to longer term population change. Similarly, there are concerns that the presence of wind farms may displace seabirds from preferred foraging areas [24], but there is limited understanding either of the extent of such effects or of the individual and population consequences of displacement, should it occur.

For other management issues, such as bycatch, estimates of Potential Biological Removal have provided management limits for human-caused mortality in mammals [70,71], but this approach cannot be used for assessing non-lethal impacts. To address this, a framework was developed called "Population Consequences of Acoustic Disturbance" (PCAD) [29,72]. The aim of this approach is to link behavioral responses by individuals and their vital rates to determine the consequences for the population. In addition to the spatially-explicit information on distribution and abundance typically collected for impact assessments, this approach also requires knowledge of dose–response relationships to link behavioral responses and demographic parameters. Since a general characterization of the dose–response relationship between received noise levels and changes in vital rates does not exist for marine mammals, expert judgment has been used to link individual impacts to changes in survival or reproductive rates [73].

Given the uncertainties involved, the population level assessments required from developers by U.K. regulators have been very conservative, and are expected to overestimate the impacts to populations. Nevertheless, the application of such an approach to a harbor seal population suggested that the population trends were largely driven by the baseline dynamics of the population and, even in a worst-case scenario of impacts, only a short term reduction in numbers would be expected to occur [58]. The long-term dynamics appeared relatively robust to uncertainty in key assumptions, but there is still a strong reliance on expert judgment and many assumptions are made. Focused studies around subsequent developments are now required to test these modeling assumptions and frameworks to ensure they are robust and, if appropriate, made less conservative in the future.

There is also a strong reliance on expert judgment in seabird collision risk models and sensitivity indices [24,60]. Avoidance rates are applied to collision risk models, but for many species they are not based on empirical data. Work is ongoing to provide estimates of these, but has been hampered by a lack of suitable techniques. The importance of this human-induced mortality on seabirds may depend on the current status of the population, with conservation concerns potentially being greater for populations that are currently in decline. For example, black-legged kittiwakes (*Rissa tridactyla*) have declined by more than 50% since 1990 in the North Sea [74]. The cause of this decline has mainly been attributed to poor breeding success as a result of reduced recruitment of their prey species, the lesser sandeel (*Ammodytes marinus*), linked to warm winters and the presence of a local sandeel fishery [74]. Black-legged kittiwakes generally fly below the minimum height of any turbine's rotor blades, but there were approximately 15.7% of flights that occurred within a generic collision risk height band defined as 20–150 m above sea level [62] and their avoidance response is unknown. The potential additional mortality that offshore wind farms could induce for this declining population makes this species of particular concern in the environmental assessment and consenting process.

Seabirds are considered at their most vulnerable when wind energy sites are proposed near their breeding colonies. During the breeding season, they make regular trips between their nest and foraging grounds. This could reduce the collision risk for wind farms proposed further offshore, but there is generally less known about the distribution and habitat use of seabirds in these areas outside of the breeding season, and their connectivity with any protected areas. As wind farms move further offshore such knowledge gaps will need to addressed.

### 6.1.5 MEASURING RESPONSES

A BACI (before-after-control-impact) design was initially recommended to assess the responses of marine mammals to wind farm construction and operation [54]. However, this type of design is limited in its ability to characterize spatial variability, assigning samples to only a treatment or control strata [75]. There are also arbitrary requirements for the selection of control sites, which include being far enough away to be unaffected by the potential disturbance, but close enough that the areas are comparable. Some stressors have a large area of potential effect, which makes it difficult to identify suitable control sites with similar ecological characteristics. Differences in variability between sites can also be a problem in statistically detecting impacts [76]. The BACI design is appropriate where there are defined boundaries for the impacted areas, but a gradient design will be more sensitive to change when a contaminant or sound disperses with distance from a point source. A gradient design requires classifying samples according to distance and removes the issue of selecting a control site. It is also more powerful than a randomized Control Impact design at detecting changes due to disturbance [77]. It has recently been demonstrated to be more effective in terms of studying spatial displacement of harbor porpoises in response to pile-driving and detecting how temporal effects differ with distance [48,78]. Furthermore, whilst BACI designs provide opportunities to identify whether or not impacts have occurred, gradient designs can also be used to assess the spatial scale of any impacts, thus informing future spatial planning decisions.

Data collection techniques used for characterizing a site in the planning stages may not be the most appropriate tools for assessing impacts. Visual surveys for both birds and mammals have generally been used to describe their abundance and distribution in planning applications. These techniques are unlikely to have enough power to detect changes in behavior or fine scale spatial or temporal shifts in distribution, since observers can only be in one place at a time and can only reliably survey in calm sea conditions during daylight hours. Our research has shown that acoustic methods for assessing impacts to marine mammals have much greater power to detect change [49,55], and techniques such as GPS tracking, radar, and fixed cameras are likely to provide more useful data for seabirds [64]. GPS tracking has been used for many species and provides high resolution data. In a recent study, it revealed harbor seals foraging around wind turbines in the North Sea [11]. Acoustic telemetry has been a valuable tool for tracking fish [79] and also turtle movements [80], and the technology now exists to use this to examine the long-term, fine-scale movements of aquatic animals [81].

The presence of other disturbance sources unrelated to the wind farm activity may compromise efforts to compare periods before and after construction events. For example, during our study of the impacts of the Beatrice Demonstrator Wind Farm project, we later discovered that hydrographic and seismic surveys had also been conducted in the area during the construction period [49]. Determining the cause of any observed effects is therefore confounded by these additional activities. Communication and coordination amongst planners, regulators, industry and scientists is therefore essential to ensure that impact studies can be properly designed, and any cumulative impacts caused by multiple events during construction or by other human activities can be taken into account [14,82]. Greater involvement of species group specialists during the planning process and engineering design phases may also help to minimize any environmental conflicts at an early stage. Careful spatial planning of wind farms has been identified as a key factor for profitability and environmental protection [83]. Consideration of the increased development of local ports to support construction and maintenance of offshore wind farms and the consequent environmental impacts is also important.

### 6.1.6 LEARNING FROM OTHER INDUSTRIES

There are three existing industries whose experiences have been usefully applied to environmental research surrounding offshore wind energy developments in European waters. These are onshore wind farms, seismic surveys for oil and gas exploration, and floating oil platforms. Our knowledge of bird vulnerability and mortality from wind farms has largely been based on those on land. Direct measurements of mortality from offshore wind farms are much more difficult because of the difficulty of finding corpses at sea. The lack of direct measurements of flight height distributions and avoidance responses for many seabird species means there is still considerable uncertainty in the mortality estimates and the consequent energetic costs of avoidance behaviors for offshore wind farms, but the modeling approaches developed for terrestrial wind farms have provided a robust framework to begin these assessments [84,85].

Airguns used in seismic surveys for oil and gas exploration produce loud multiple pulsed noises with energy mainly below 1 kHz, but also extending to much higher frequencies [55,86]. Pile-driving also produces loud multiple-pulsed sounds, although they tend to be more broadband with the major amplitude at 100–500 Hz compared to a seismic airgun array at 10–120 Hz [87]. In addition, pile-driving has a shorter interval between pulses at about one second [47,88] as opposed to seismic airgun surveys at 10 seconds or more [87]. Changes in distribution and vocal behavior by marine mammals [55,89-92], and diving behavior by loggerhead turtles [93] have been observed to occur in response to seismic surveys. Following environmental concerns about the impact of these explosive sounds, underwater sound propagation models have been developed to estimate received levels. These are used to determine the distance at which injury or disturbance may occur and to develop mitigation and monitoring plans to reduce noise exposure [94,95]. These approaches have often been adopted during assessments and mitigation of pile-driving activity at offshore wind farms.

Mitigation measures for marine mammals during seismic surveys typically include a soft start or ramp-up to gradually increase the intensity of an airgun array up to full power over a period of 20 minutes or more. This approach is to allow sufficient time for animals to leave the immediate vicinity and avoid harmful noise levels. Similar approaches have been applied to the blow energy intensity during pile-driving. However, whilst this mitigation measure is implemented as a 'common sense' approach, no studies have yet investigated its effectiveness systematically [94]. The form, probability and extent of a marine mammal's response to anthropogenic sound will be affected by a variety of factors. Animals may have different tolerances for increasing sound levels depending on their current behavior, experience, motivation and conditioning [53]. One study observed an avoidance response away from the ramp-up of a 2-D seismic survey by a subgroup of pilot whales (*Globicephala macrorhynchus*) that began when they were 750 m from the airgun array [96], but interpreting the reactions of animals can be difficult because responses can be vertical and/or horizontal. There is a need for further research to assess the efficacy of the ramp-up soft start procedure for mitigating effects on marine mammals.

Another mitigation measure typically used is the monitoring of an exclusion zone. Marine mammal observers are required to visually, and sometimes also acoustically, monitor within a zone in close proximity to the source to ensure the absence of marine mammals (and possibly other protected species such as sea turtles) before beginning piling e.g. [97].

This zone may be a pre-defined fixed distance from the source or based on the expected sound levels. However, there is generally a mismatch between the relatively small area monitored for animals and the potential area of impact, which is likely to be considerably larger [98]. The exclusion zone is aimed at reducing near-field noise exposure and protecting animals from direct physical harm.

Visual observations will be limited during poor visibility conditions and for deep-diving species, such as beaked whales. It is also recognized that this is unlikely to be effective in mitigating behavioral responses over greater distances and that disturbance in the far-field is still likely to occur [55,86,99]. The use of real-time technologies, such as passive acoustic monitoring [100], may be a cost-effective approach to achieve detection coverage over a much larger area for vocalizing animals. Detailed studies to estimate received levels at various distances should be conducted during the planning stages to take into account variations in sound propagation among locations and use this, together with spatiotemporal information on marine mammal occurrence, to identify priority areas for monitoring and mitigation. Current mitigation plans also do not consider the impacts on marine mammal prey species. It should be identified whether any prey species (e.g. fish, squid) are potentially sensitive to noise and disturbance and considered in management plans accordingly to avoid secondary, trophic-level effects, as well as impacts to the fishing industry [18,98]. Efforts are underway to develop technologies to reduce source levels and noise propagation around offshore wind farm sites to help minimize biological impacts e.g. [101].

One measure that could reduce or eliminate the need for pile-driving is the development of floating wind turbine technologies, which are now being considered for deep water (>50 m) sites [2,4,102]. Concerns have been raised over possible entanglement risk in the moorings used to secure the platform to anchors on the seabed. However, the risk would appear to be small as the cables will be under tension and such moorings would be very similar to those widely used for floating oil platforms. Assessments of interactions with wildlife and existing floating oil platforms could therefore inform risk assessments for floating offshore wind turbines and identify what species or groups, if any, may be vulnerable to entanglement.

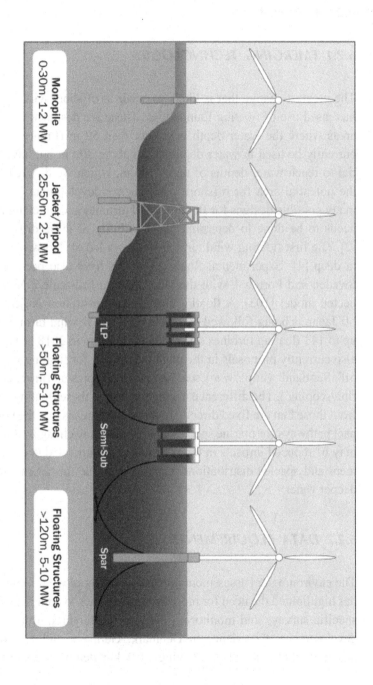

**FIGURE 4:** Types of offshore wind turbine foundations (reproduced from ref. 102, source Principle Power). Monopile and tripod/jacket foundations are currently proven technologies. Floating structures have been using three main types of foundations, which are adapted from the oil and gas industry: the Tension Leg Platform (TLP), semi-submersible (Semi-sub), and Spar Buoy (Spar).

## 6.2 THE FUTURE

### 6.2.1 EMERGING TECHNOLOGIES

The greatest change that is likely to occur in offshore wind energy is the increased use of floating foundations. These are designed for deep water areas where the water depth is greater than 50 m (Figure 4). They can currently be used in water depths up to about 300 m but have the potential to reach water depths of up to 700 m, which would greatly increase the potential area for offshore wind energy development [4]. There are many possible designs for floating wind turbines and much more research needs to be done to determine the feasibility of these different options [2]. The first floating wind turbine was installed off Norway in water 220 m deep [4]. Experimental floating turbines have also been installed off Sweden and Portugal, with the latter being a full-scale 2 MW grid connected model [103]. A floating turbine demonstration project of 2 MW off Japan is being followed by a plan for a 1 GW wind farm consisting of up to 143 floating turbines scheduled for start-up in 2018 [104]. There are also currently proposals in the planning system for floating wind turbines off Scotland (http://www.scotland.gov.uk/Topics/marine/Licensing/marine/scoping ). The difference in construction of these floating foundations from those that are fixed directly to the seabed means that the potential impact pathways for marine species and habitats may change. Although there may be reduced impacts in terms of noise, our knowledge of the environment and species distributions tends to decrease further offshore and in deeper water.

### 6.2.2 DATA REQUIREMENTS

The environmental assessment process for offshore wind farms in Europe has highlighted the need for more synoptic studies to complement the site-specific surveys and monitoring that may be required around particular developments. Experience in Denmark, Germany and the Netherlands has highlighted the value of having a few key demonstrator sites to study interactions with key receptor species in these shallow North Sea areas e.g. [9,48,78]. Development of a broader suite of demonstrator sites is now required to understand potential interactions with a wider range of species and habitats that will result from the expansion of this industry. Such demonstrator sites could be focused on areas that build on existing research programs or where there are specific species of concern so that parameters of interest can be determined and models for assessing impacts developed and tested. Where regulators are required to consider potential population level effects on protected species, demonstrator sites should be selected to maximize the opportunity for linkage with individual based demographic studies [105]. This approach offers the potential to explore whether individual and colony specific variation in exposure to stressors such as noise or collision risk influences reproduction and survival rates. Critically, individual based studies can also be used to assess how the impacts of particular stressors interact with broader scale variation in environmental conditions e.g. [106] or vary over time e.g. [107]. For example, long-term data collection on bottlenose dolphins and harbor seals in the Moray Firth, Northeast Scotland, means that data on demography and fecundity are available as a baseline and can be used to determine if there are any changes in these vital rates during construction activities [58,108]. Focused studies such as these will be especially important for developing and testing the modeling frameworks that have been used to assess the impacts of construction noise [58] or collision risk [65], supporting a more general understanding of the longer-term population consequence of the short-term interactions recognized in current assessments. Information on these ecological processes that can only be obtained from a few focused studies at demonstrator sites can then be integrated with site specific data on distribution and abundance at proposed wind energy sites. This will provide more robust assessments of the population consequences of future developments.

## 6.3 CONCLUSIONS

As offshore wind farms grow in size and number around the world, several changes in the priorities for environmental research and assessments are occurring. Firstly, there are an increasing number of cases where more than one wind farm project may occur within the home range of a population. Consequently, cumulative impact assessments, which should be made at the population level, will become increasingly important when assessing the effect of these activities on marine species and populations. Secondly, for species such as marine mammals, it is becoming increasingly clear that the most significant consequences of offshore wind farm construction and operation are likely to occur as a result of avoidance of construction noise or structures rather than direct mortality. Hence there needs to be a greater focus on assessing the longer-term impact of any behavioral responses through changes in energetic costs, survival or fecundity. Finally, as offshore wind farms increase in scale, there is a need to put any observed biological impacts into a population context. This requires an understanding of the relative scale of any impacts in relation to existing natural variation and other anthropogenic drivers such as fisheries bycatch or exploitation. Only then can the population consequences be modeled and conservation priorities be identified.

Drewitt and Langston [109] previously recommended a number of best practice measures for reducing the impacts of wind farms on birds. These recommendations included ensuring that key areas of conservation importance and sensitivity are avoided, grouping turbines to avoid alignment perpendicular to main flight paths or migration corridors, timing construction to avoid sensitive periods, and timing and routing maintenance trips to reduce disturbance from boats, helicopters and personnel. In practice, it is unlikely that all of these recommendations can be met given the challenge of balancing the needs of all stakeholders in the marine spatial planning process. In particular, many of the sites suitable for offshore wind energy development, such as offshore sandbanks, are also important habitats for marine species and fisheries. There is therefore a need for careful consideration in finer scale spatial planning and the identification of other mitigation measures to minimize environmental and human user conflicts. Strategic and targeted research around the next generation of offshore wind farm sites is now required to support the regulators' need to achieve a balance between climate change targets and existing environmental legislation.

# REFERENCES

1. Toke D: The UK offshore wind power programme: a sea-change in UK energy policy? Energy Policy 2011, 39:526-534.
2. Breton SP, Moe G: Status, plans and technologies for offshore wind turbines in Europe and North America. Renew Energy 2009, 34:646-654.
3. European Wind Energy Association: The European Offshore Wind Industry - Key Trends and Statistics 2013. Brussels, Belgium: A report by the European Wind Energy Association; 2014.
4. Sun X, Huang D, Wu G: The current state of offshore wind energy technology development. Energy 2012, 41:298-312.
5. Inger R, Attrill MJ, Bearhop S, Broderick AC, Grecian WJ, Hodgson DJ, Mills C, Sheehan E, Votier SC, Witt MJ, Godley BJ: Marine renewable energy: potential benefits to biodiversity? An urgent call for research. J Appl Ecol 2009, 46:1145-1153.
6. Boehlert GW, Gill AB: Environmental and ecological effects of ocean renewable energy development: a current synthesis. Oceanography 2010, 23:68-81.
7. Gill AB: Offshore renewable energy: ecological implications of generating electricity in the coastal zone. J Appl Ecol 2005, 42:605-615.
8. Wilhelmsson D, Malm T, Öhman MC: The influence of offshore windpower on demersal fish. ICES J Mar Sci 2006, 63:775-784.
9. Lindeboom HJ, Kouwenhoven HJ, Bergman MJN, Bouma S, Brasseur S, Daan R, Fijn RC, De Haan D, Dirksen S, van Hal R, Hille Ris Lambers R, ter Hofstede R, Krijgsveld KL, Leopold M, Scheidat M: Short-term ecological effects of an offshore wind farm in the Dutch coastal zone; a compilation. Environ Res Lett 2011, 6:035101.
10. Maar M, Bolding K, Petersen JK, Hansen JLS, Timmermann K: Local effects of blue mussels around turbine foundations in an ecosystem model of Nysted off-shore wind farm, Denmark. J Sea Res 2009, 62:159-174.
11. Russell DJF, Brasseur SMJM, Thompson D, Hastie GD, Janik VM, Aarts G, McClintock BT, Matthiopoulos J, Moss SEW, McConnell B: Marine mammals trace anthropogenic structures at sea. Curr Biol 2014, 24:R638-R639.
12. Buck BH, Krause G, Rosenthal H: Extensive open ocean aquaculture development within wind farms in Germany: the prospect of offshore co-management and legal constraints. Ocean Coast Manag 2004, 47:95-122.
13. Renewable UK: Cumulative impact assessment guidelines: Guiding principles for cumulative impacts assessment in offshore wind farms. 2013. Available at: http://www.renewableuk.com/en/publications/index.cfm/cumulative-impact-assessment-guidelines
14. Masden EA, Fox AD, Furness RW, Bullman R, Haydon DT: Cumulative impact assessments and bird/wind farm interactions: Developing a conceptual framework. Environ Impact Assess Rev 2010, 30:1-7.
15. Dolman S, Simmonds M: Towards best environmental practice for cetacean conservation in developing Scotland's marine renewable energy. Mar Policy 2010, 34:1021-1027.

16. Madsen PT, Wahlberg M, Tougaard J, Lucke K, Tyack P: Wind turbine underwater noise and marine mammals: implications of current knowledge and data needs. Mar Ecol Prog Ser 2006, 309:279-295.

17. Thomsen F, Lüdemann K, Kafemann R, Piper W: Effects of offshore wind farm noise on marine mammals and fish. Biola, Hamburg: Germany on behalf of COW-RIE Ltd.; 2006.

18. Popper AN, Hastings MC: The effects of human-generated sound on fish. Integr Zool 2009, 4:43-52.

19. Gill AB, Bartlett M, Thomsen F: Potential interactions between diadromous fishes of U.K. conservation importance and the electromagnetic fields and subsea noise from marine renewable energy developments. J Fish Biol 2012, 81:664-695.

20. Popper AN, Hastings MC: The effects of anthropogenic sources of sound on fishes. J Fish Biol 2009, 75:455-489.

21. Tougaard J, Henriksen OD, Miller LA: Underwater noise from three types of offshore wind turbines: Estimation of impact zones for harbor porpoises and harbor seals. J Acoust Soc Am 2009, 125:3766-3773.

22. Marmo B, Roberts I, Buckingham MP, King S, Booth C: Modelling of noise effects of operational offshore wind turbines including noise transmission through various foundation types. Edinburgh: Scottish Government; 2013.

23. Desholm M, Kahlert J: Avian collision risk at an offshore wind farm. Biol Lett 2005, 1:296-298.

24. Furness RW, Wade HM, Masden EA: Assessing vulnerability of marine bird populations to offshore wind farms. J Environ Manag 2013, 119:56-66.

25. Tricas T, Gill A, Normandeau, Exponent: Effects of EMFs from undersea power cables on elasmobranchs and other marine species. Camarillo, CA: U.S. Dept. of the Interior, Bureau of Ocean Energy Management, Regulation, and Enforcement, Pacific OCS Region; 2011. OCS Study BOEMRE 2011–09

26. Gill AB, Huang Y, Gloyne-Philips I, Metcalfe J, Quayle V, Spencer J, Wearmouth V: COWRIE 2.0 Electromagnetic Fields (EMF) Phase 2: EMF-sensitive fish response to EM emmisions from sub-sea electricity cables of the type used by the offshore renewable energy industry. Thetford, UK: Commissioned by COWRIE Ltd (project reference COWRIE-EMF-1-06); 2009.

27. Westerberg H, Lagenfelt I: Sub-sea power cables and the migration behaviour of the European eel. Fish Manag Ecol 2008, 15:369-375.

28. Southall BL, Bowles AE, Ellison WT, Finneran JJ, Gentry RL, Greene CR Jr, Kastak D, Ketten DR, Miler JH, Nachtigall PE, Richardson WJ, Thomas JA, Tyack PL: Marine mammal noise exposure criteria: initial scientific recommendations. Aquat Mamm 2007, 33:411-521.

29. National Research Council: Marine Mammal Populations and Ocean Noise: Determining When Ocean Noise Causes Biologically Significant Effects. Washington, DC: National Academy Press; 2005.

30. Wahlberg M, Westerberg H: Hearing in fish and their reactions to sounds from offshore wind farms. Mar Ecol Prog Ser 2005, 288:295-309.

31. Hawkins AD, Popper AN: Assessing the impacts of underwater sounds on fishes and other forms of marine life. Acoust Today 2014, 10:30-41.

32. Casper BM, Halvorsen MB, Matthews F, Carlson TJ, Popper AN: Recovery of baro-trauma injuries resulting from exposure to pile driving sound in two sizes of hybrid striped bass. PLoS ONE 2013, 8:e73844.

33. Casper BM, Popper AN, Matthews F, Carlson TJ, Halvorsen MB: Recovery of baro-trauma injuries in Chinook salmon, Oncorhynchus tshawytscha from exposure to pile driving sound. PLoS ONE 2012, 7:e39593.

34. Halvorsen MB, Casper BM, Matthews F, Carlson TJ, Popper AN: Effects of expo-sure to pile-driving sounds on the lake sturgeon, Nile tilapia and hogchoker. Proc R Soc Lond B Biol Sci 2012, 279:4705-4714.

35. Bolle LJ, de Jong CAF, Bierman SM, Van Beek PJG, Van Keeken OA, Wessels PW, Van Damme CJG, Winter HV, De Haan D, Dekeling RPA: Common sole larvae survive high levels of pile-driving sound in controlled exposure experiments. PLoS ONE 2012, 7:e33052.

36. Thomsen F, Mueller-Blenkle C, Gill A, Metcalfe J, McGregor PK, Bendall V, An-dersson MH, Sigray P, Wood D: Effects of pile driving on the behavior of cod and sole. In The Effects of Noise on Aquatic Life. Edited by Popper AN, Hawkins A. New York, USA: Springer; 2012:387-388. [Advances in Experimental Medicine and Biology, Volume 730]

37. Mueller-Blenkle C, Gill AB, McGregor PK, Andersson MH, Sigray P, Bendall V, Metcalfe J, Thomsen F: A novel field study setup to investigate the behavior of fish related to sound. In The Effects of Noise on Aquatic Life. Edited by Popper AN, Hawkins A. New York, USA: Springer; 2012:389-391. [Advances in Experimental Medicine and Biology, Volume 730]

38. Kaiser MJ, Galanidi M, Showler DA, Elliott AJ, Caldow RWG, Rees EIS, Stillman RA, Sutherland WJ: Distribution and behaviour of Common Scoter Melanitta nigra relative to prey resources and environmental parameters. Ibis 2006, s1:110-128.

39. Pelletier SK, Omland K, Watrous KS, Peterson TS: Information Synthesis on the Potential for Bat Interactions with Offshore Wind Facilities - Final Report. Herndon, VA: U.S. Department of the Interior, Bureau of Ocean Energy Management, Head-quarters; 2013. OCS Study BOEM 2013–01163

40. Sjollema AL, Gates JE, Hilderbrand RH, Sherwell J: Offshore activity of bats along the Mid-Atlantic Coast. Northeast Nat 2014, 21:154-163.

41. Kunz TH, Arnett EB, Erickson WP, Hoar AR, Johnson GD, Larkin RP, Strickland MD, Thresher RW, Tuttle MD: Ecological impacts of wind energy development on bats: questions, research needs, and hypotheses. Front Ecol Environ 2007, 5:315-324.

42. Waring GT, Wood SA, Josephson E: Literature search and data synthesis for marine mammals and sea turtles in the U.S. Atlantic from Maine to the Florida Keys. New Orleans, LA: U.S Department of the Interior, Bureau of Ocean Energy Management, Gulf of Mexico OCS Region; 2012. OCS Study BOEM 2012–109

43. Dow Piniak WE, Eckert SA, Harms CA, Stringer EM: Underwater Hearing Sensi-tivity of The Leatherback Sea Turtle (Dermochelys coriacea): Assessing The Poten-tial Effect of Anthropogenic Noise. Herndon, VA: U.S Department of the Interior, Bureau of Ocean Energy Management, Headquarters; 2012. OCS Study BOEM 2012–01156

44. Rein CG, Lundin AS, Wilson SJK, Kimbrell E: Offshore Wind Energy Development Site Assessment and Characterization: Evaluation of the Current Status and Euro-

pean Experience. Herndon, VA: U.S. Department of the Interior, Bureau of Ocean Energy Management, Office of Renewable Energy Programs; 2013. OCS Study BOEM 2013–0010

45. Carstensen J, Henriksen OD, Teilmann J: Impacts of offshore wind farm construction on harbour porpoises: acoustic monitoring of echolocation activity using porpoise detectors (T-PODs). Mar Ecol Prog Ser 2006, 321:295-308.

46. Tougaard J, Carstensen J, Teilmann J: Pile driving zone of responsiveness extends beyond 20 km for harbor porpoises (Phocoena phocoena (L.)). J Acoust Soc Am 2009, 126:11-14.

47. Bailey H, Senior B, Simmons D, Rusin J, Picken G, Thompson PM: Assessing underwater noise levels during pile-driving at an offshore windfarm and its potential impact on marine mammals. Mar Pollut Bull 2010, 60:888-897.

48. Brandt MJ, Diederichs A, Betke K, Nehls G: Responses of harbour porpoises to pile driving at the Horns Rev II offshore wind farm in the Danish North Sea. Mar Ecol Prog Ser 2011, 421:205-216.

49. Thompson PM, Lusseau D, Barton T, Simmons D, Rusin J, Bailey H: Assessing the responses of coastal cetaceans to the construction of offshore wind turbines. Mar Pollut Bull 2010, 60:1200-1208.

50. Kastelein RA, Van Heerden D, Gransier R, Hoek L: Behavioral responses of a harbor porpoise (Phocoena phocoena) to playbacks of broadband pile driving sounds. Mar Environ Res 2013, 92:206-214.

51. Koschinski S, Culik BM, Henriksen OD, Tregenza N, Ellis G, Jansen C, Kathe G: Behavioural reactions of free-ranging porpoises and seals to the noise of a simulated 2 MW windpower generator. Mar Ecol Prog Ser 2003, 265:263-273.

52. Degraer S, Brabant R, Rumes B: Environmental impacts of offshore wind farms in the Belgian part of the North Sea: Learning from the past to optimise future monitoring programmes: 26–28 November 2013. Brussels, Belgium: Royal Belgian Institute of Natural Sciences; 2013.

53. Ellison WT, Southall BL, Clark CW, Frankel AS: A new context-based approach to assess marine mammal behavioral responses to anthropogenic sounds. Conserv Biol 2012, 26:21-28.

54. Diederichs A, Nehls G, Dähne M, Adler S, Koschinski S, Verfuß U: Methodologies for measuring and assessing potential changes in marine mammal behaviour, abundance or distribution arising from the construction, operation and decommissioning of offshore windfarms. Germany: BioConsult SH report to COWRIE Ltd; 2008.

55. Thompson PM, Brookes KL, Graham IM, Barton TR, Needham K, Bradbury G, Merchant ND: Short-term disturbance by a commercial two-dimensional seismic survey does not lead to long-term displacement of harbour porpoises. Proc R Soc Lond B Biol Sci 2013, 280:20132001.

56. Bailey H, Hammond PS, Thompson PM: Modelling harbour seal habitat by combining data from multiple tracking systems. J Exp Mar Biol Ecol 2014, 450:30-39.

57. Bailey H, Clay G, Coates EA, Lusseau D, Senior B, Thompson PM: Using T-PODs to assess variations in the occurrence of coastal bottlenose dolphins and harbour porpoises. Aquat Conserv Mar Freshwat Ecosyst 2010, 20:150-158.

58. Thompson PM, Hastie GD, Nedwell J, Barham R, Brookes KL, Cordes LS, Bailey H, McLean N: Framework for assessing impacts of pile-driving noise from offshore

wind farm construction on a harbour seal population. Environ Impact Assess Rev 2013, 43:73-85.

59. Barrios L, Rodríguez A: Behavioural and environmental correlates of soaring-bird mortality at on-shore wind turbines. J Appl Ecol 2004, 41:72-81.

60. Garthe S, Hüppop O: Scaling possible adverse effects of marine wind farms on sea-birds: developing and applying a vulnerability index. J Appl Ecol 2004, 41:724-734.

61. Camphuysen KCJ, Fox TAD, Leopold MMF, Petersen IK: Towards standardised seabirds at sea census techniques in connection with environmental impact assessments for offshore wind farms in the U.K.. London: Crown Estate Commissioners; 2004. [Report by Royal Netherlands Institute for Sea Research and the Danish National Environmental Research Institute to COWRIE BAM 02–2002]

62. Cook ASCP, Johnston A, Wright LJ, Burton NHK: A Review of Flight Heights and Avoidance Rates of Birds in Relation to Offshore Wind Farms. Norfolk, UK: British Trust for Ornithology on behalf of The Crown Estate, Project SOSS-02; 2012. BTO Research Report Number 618

63. Johnston A, Cook ASCP, Wright LJ, Humphreys EM, Burton NHK: Modelling flight heights of marine birds to more accurately assess collision risk with offshore wind turbines. J Appl Ecol 2014, 51:31-41.

64. Plonczkier P, Simms IC: Radar monitoring of migrating pink-footed geese: behavioural responses to offshore wind farm development. J Appl Ecol 2012, 49:1187-1194.

65. Chamberlain DE, Rehfisch MR, Fox AD, Desholm M, Anthony SJ: The effect of avoidance rates on bird mortality predictions made by wind turbine collision risk models. Ibis 2006, 148:198-202.

66. Masden EA, Haydon DT, Fox AD, Furness RW: Barriers to movement: Modelling energetic costs of avoiding marine wind farms amongst breeding seabirds. Mar Pollut Bull 2010, 60:1085-1091.

67. Busch M, Kannen A, Garthe S, Jessopp M: Consequences of a cumulative perspective on marine environmental impacts: Offshore wind farming and seabirds at North Sea scale in context of the EU Marine Strategy Framework Directive. Ocean Coast Manag 2013, 71:213-224.

68. McCann J: Developing Environmental Protocols and Modeling Tools to Support Ocean Renewable Energy and Stewardship. Herndon, VA: U.S. Department of the Interior, Bureau of Ocean Energy Management, Office of Renewable Energy Programs; 2012. OCS Study BOEM 2012–082

69. Teilmann J, Carstensen J: Negative long term effects on harbour porpoises from a large scale offshore wind farm in the Baltic - evidence of slow recovery. Environ Res Lett 2012, 7:045101.

70. Wade PR: Calculating limits to the allowable human-caused mortality of cetaceans and pinnipeds. Mar Mammal Sci 1998, 14:1-37.

71. Butler JRA, Middlemas SJ, McKelvey SA, McMyn I, Leyshon B, Walker I, Thompson PM, Boyd IL, Duck C, Armstrong JD, Graham IM, Baxter JM: The Moray Firth Seal Management Plan: an adaptive framework for balancing the conservation of seals, salmon, fisheries and wildlife tourism in the UK. Aquat Conserv Mar Freshwat Ecosyst 2008, 18:1025-1038.

72. New LF, Clark JS, Costa DP, Fleishman E, Hindell MA, Klanjšček T, Lusseau D, Kraus S, McMahon CR, Robinson PW, Schick RS, Schwarz LK, Simmons SE, Thomas L, Tyack P, Harwood J: Using short-term measures of behaviour to estimate long-term fitness of southern elephant seals. Mar Ecol Prog Ser 2014, 496:99-108.

73. Harwood J, King S, Schick R, Donovan C, Booth C: A protocol for implementing the interim population consequences of disturbance (PCoD) approach: Quantifying and assessing the effects of UK offshore renewable energy developmenets on marine mammal populations: Report number SMRUL-TCE-2013-014. Scott Mar Freshwater Sci 2014, 5:2.

74. Frederiksen M, Wanless S, Harris MP, Rothery P, Wilson LJ: The role of industrial fisheries and oceanographic change in the decline of North Sea black-legged kittiwakes. J Appl Ecol 2004, 41:1129-1139.

75. Underwood AJ: On beyond BACI: Sampling designs that might reliably detect environmental disturbances. Ecol Appl 1994, 4:3-15.

76. Hewitt JE, Thrush SE, Cummings VJ: Assessing environmental impacts: Effects of spatial and temporal variability at likely impact scales. Ecol Appl 2001, 11:1502-1516.

77. Ellis JI, Schneider DC: Evaluation of a gradient sampling design for environmental impact assessment. Environ Monit Assess 1997, 48:157-172.

78. Dähne M, Gilles A, Lucke K, Peschko V, Adler S, Krügel K, Sundermeyer J, Siebert U: Effects of pile-driving on harbour porpoises (Phocoena phocoena) at the first offshore wind farm in Germany. Environ Res Lett 2013, 8:025002.

79. Heupel MR, Semmens JM, Hobday AJ: Automated acoustic tracking of aquatic animals: scales, design and deployment of listening station arrays. Mar Freshw Res 2006, 57:1-13.

80. Scales KL, Lewis JA, Lewis JP, Castellanos D, Godley BJ, Graham RT: Insights into habitat utilisation of the hawksbill turtle, Eretmochelys imbricata (Linnaeus, 1766), using acoustic telemetry. J Exp Mar Biol Ecol 2011, 407:122-129.

81. Espinoza M, Farrugia TJ, Webber DM, Smith F, Lowe CG: Testing a new acoustic telemetry technique to quantify long-term, fine-scale movements of aquatic animals. Fish Res 2011, 108:364-371.

82. Maxwell SM, Hazen EL, Bograd SJ, Halpern BS, Breed GA, Nickel B, Teutschel NM, Crowder LB, Benson S, Dutton PH, Bailey H, Kappes MA, Kuhn CE, Weise MJ, Mate B, Shaffer SA, Hassrick JL, Henry RW, Irvine L, McDonald BI, Robinson PW, Block BA, Costa DP: Cumulative human impacts on marine predators. Nat Commun 2013, 4:2688.

83. Punt MJ, Groeneveld RA, Van Ierland EC, Stel JH: Spatial planning of offshore wind farms: A windfall to marine environmental protection? Ecol Econ 2009, 69:93-103.

84. Band W, Madders M, Whitfield DP: Developing field and analytical methods to assess avian collision risk at wind farms. In Birds and Wind Power. Edited by De Lucas M, Janss G, Ferrer M. Barcelona, Spain: Lynx Edicions; 2005.

85. Band B, Band B: Using a collision risk model to assess bird collision risks for offshore windfarms. Norway: SOSS report for The Crown Estate; 2012.

86. Gordon J, Gillespie D, Potter J, Frantzis A, Simmonds MP, Swift R, Thompson D: A review of the effects of seismic surveys on marine mammals. Mar Technol Soc J 2003, 37:16-34.

87. OSPAR: Overview of the impacts of anthropogenic underwater sound in the marine environment. North-East Atlantic: OSPAR Convention for the Protection of the Marine Environment of the North-East Atlantic; 2009. http://www.ospar.org

88. Nedwell JR, Parvin SJ, Edwards B, Workman R, Brooker AG, Kynoch JE, Nedwell JR, Parvin SJ, Edwards B, Workman R, Brooker AG, Kynoch JE: Measurement and interpretation of underwater noise during construction and operation of offshore windfarms in UK waters. Subacoustech Report No. 544R0738 to COWRIE Ltd 2007. 978-0-9554279-5-4

89. Di Iorio L, Clark CW: Exposure to seismic survey alters blue whale acoustic communication. Biol Lett 2010, 6:51-54.

90. Castellote M, Clark CW, Lammers MO: Acoustic and behavioural changes by fin whales (Balaenoptera physalus) in response to shipping and airgun noise. Biol Conserv 2012, 147:115-122.

91. Blackwell SB, Nations CS, McDonald TL, Greene CR, Thode AM, Guerra M, Macrander AM: Effects of airgun sounds on bowhead whale calling rates in the Alaskan Beaufort Sea. Mar Mammal Sci 2013, 29:E342-E365.

92. Harris RE, Miller GW, Richardson WJ: Seal responses to airgun sounds during summer seismic surveys in the Alaskan Beaufort Sea. Mar Mammal Sci 2001, 17:795-812.

93. DeRuiter SL, Doukara KL: Loggerhead turtles dive in response to airgun sound exposure. Endanger Species Res 2012, 16:55-63.

94. Rutenko AN, Borisov SV, Gritsenko AV, Jenkerson MR: Calibrating and monitoring the western gray whale mitigation zone and estimating acoustic transmission during a 3D seismic survey, Sakhalin Island, Russia. Environ Monit Assess 2007, 134:21-44.

95. Nowacek DP, Vedenev A, Southall BL, Racca R: Development and implementation of criteria for exposure of western gray whales to oil and gas industry noise. In The Effects of Noise on Aquatic Life. Edited by Popper AN, Hawkins A. New York, USA: Springer; 2012:523-528. [Advances in Experimental Medicine and Biology, Volume 730]

96. Weir CR: Short-finned pilot whales (Globicephala macrorhynchus) respond to an airgun ramp-up procedure off Gabon. Aquat Mamm 2008, 34:349-354.

97. JNCC: Statutory Nature Conservation Agency Protocol for Minimising the Risk of Injury to Marine Mammals from Piling Noise. Aberdeen, UK: Joint Nature Conservation Committee; 2010.

98. Parsons ECM, Dolman SJ, Jasny M, Rose NA, Simmonds MP, Wright AJ: A critique of the UK's JNCC seismic survey guidelines for minimising acoustic disturbance to marine mammals: Best practise? Mar Pollut Bull 2009, 58:643-651.

99. Miller PJO, Johnson MP, Madsen PT, Biassoni N, Quero M, Tyack PL: Using at-sea experiments to study the effects of airguns on the foraging behavior of sperm whales in the Gulf of Mexico. Deep-Sea Res I 2009, 56:1168-1181.

100. Van Parijs SM, Clark CW, Sousa-Lima RS, Parks SE, Rankin S, Risch D, Van Opzeeland IC: Management and research applications of real-time and archival pas-

sive acoustic sensors over varying temporal and spatial scales. Mar Ecol Prog Ser 2009, 395:21-36.

101. Bellmann MA, Remmers P: Noise mitigation systems (NMS) for reducing pile driving noise: Experiences with the "big bubble curtain" relating to noise reduction. J Acoust Soc Am 2013, 134:4059.

102. European Wind Energy Association: Deep Water: The Next Step for Offshore Wind Energy. Brussels, Belgium: A report by the European Wind Energy Association; 2013.

103. European Wind Energy Association: The European Offshore Wind Industry Key 2011 Trends and Statistics. Brussels, Belgium: A report by the European Wind Energy Association; 2012.

104. Stokes I: Hotspots: Scotland and Fukushima. Renewable Energy Focus 2013, 14:10-11.

105. Clutton-Brock T, Sheldon BC: Individuals and populations: the role of long-term, individual-based studies of animals in ecology and evolutionary biology. Trends Ecol Evol 2010, 25:562-573.

106. Votier SC, Hatchwell BJ, Beckerman A, McCleery RH, Hunter FM, Pellatt J, Trinder M, Birkhead TR: Oil pollution and climate have wide-scale impacts on seabird demographics. Ecol Lett 2005, 8:1157-1164.

107. Véran S, Gimenez O, Flint E, Kendall WL, Doherty PF, Lebreton JD: Quantifying the impact of longline fisheries on adult survival in the black-footed albatross. J Appl Ecol 2007, 44:942-952.

108. New LF, Harwood J, Thomas L, Donovan C, Clark JS, Hastie G, Thompson PM, Cheney B, Scott-Hayward L, Lusseau D: Modelling the biological significance of behavioural change in coastal bottlenose dolphins in response to disturbance. Funct Ecol 2013, 27:314-322.

109. Drewitt AL, Langston RHW: Assessing the impacts of wind farms on birds. Ibis 2006, 148:29-42.

CHAPTER 7

# Socioeconomic Impacts of Wind Farm Development: A Case Study of Weatherford, Oklahoma

JOHN SCOTT GREENE AND MARK GEISKEN

## 7.1 BACKGROUND

The use of wind as an energy source in Oklahoma has a long history [1]. For example, wind energy was used by early settlers and farmers in Western Oklahoma to power well pumps. The early settlers were able to use these pumps to irrigate and make farming possible in areas where climate may have otherwise prohibited it. Similarly, today's society could choose to continue to use fossil fuels, which have significant harmful effects [2], or could choose to create a diversified energy portfolio [3]. Schiermeier et al. in 2008 [4] illustrate the potential mechanisms and impacts of 'electricity without carbon', as utilities and policymakers move toward a low carbon economy. Nationally, there have been increasing efforts to promote renewable energy as a response to the awareness of the limited supply of fossil fuels, to meet the growing energy demand, and to reduce the harmful environmental impact of fossil fuel use. Organizations such as the National Renewable Energy Laboratory, the American Wind Energy Association,

*Socioeconomic Impacts of Wind Farm Development: A Case Study of Weatherford, Oklahoma.* ©
Greene JS and Geisken M. Energy, Sustainability and Society 3,2 (2013), doi:10.1186/2192-0567-3-
2. Licensed under Creative Commons Attribution 2.0 Generic License, http://creativecommons.org/
licenses/by/2.0/.

and others have been researching and promoting renewable energy in the USA. State and local governments as well as the federal government have realized that not only can renewable energy be a way to meet future energy demands but it could also promote economic growth in rural communities. Some of these rural communities have experienced job losses and declining population in recent years [5].

At the same time the groups mentioned above have been highlighting and promoting potential impacts, wind farm developers have also been looking for potential markets and locations for expansion. Figure 1 shows the wind resources across Oklahoma (the study location is indicated by the square in the figure). The geographic distribution of this resource varies severely from east to west. In the eastern part of the state, rough terrain helps to reduce wind resources. However, western Oklahoma has large areas of commercially viable wind resources, particularly associated with a series of west–east running ridge lines across the western portion of the state. The class 3 wind areas in Figure 1 show those areas in which a wind farm could be potentially economically viable. Rural communities have been some of the hardest hit areas economically in recent decades, suffering large losses of population and jobs [5]. Figure 2 overlays the commercially viable wind in Oklahoma with those counties undergoing population and job loss. This figure shows that wind-driven economic development would have an even greater impact in these areas.

Wind farms have different impacts on local economies. They provide both short-term and long-term employment during different phases of development. Landowners also benefit in the form of annual royalty payments. Local economies will benefit greatest if the local community can provide a wide range of goods and services that can be used during the construction of the wind farms. The extent to which the local economy offers goods and services will determine how significant the ultimate impact will be on the local economy. Local ownership can also play an important role in the overall impact of the wind farm development. As stated by Phiminster and Roberts (2012), development of the wind energy sector is often listed as a way to support rural economies [6]. As they conclude 'with no local ownership, while rural GDP increases, there is almost no effect on household incomes due to the limited direct linkages of the onshore wind sector.' Similarly, while local ownership can result in a benefit to household incomes, 'there are still limited positive spill-over effects on the wider economy unless factor income is re-invested in local capital' [6].

If the local economy does not offer a wide range of goods and services, these goods and services must be brought in from elsewhere; thus, this income will leave the local economy. The income leaving the community to pay for these goods and services is referred to as 'leakage' [7,8]. Leakage occurs when the developer of the wind farm must leave the local community and contract with companies outside of the local community.

County demographics, including population levels, education levels, and amount of economic diversity, help to further identify the economic impacts of wind farm development [9]. For example, by looking at a range of factors across wind-rich counties in the central USA, Brown et al. in 2012 found that there is a median of 0.5 jobs per megawatt of wind power capacity [9]. For the study period of 2000 to 2008, this represents a 'median increase in total county personal income and employment of 0.2% and 0.4% for counties with installed wind power'.

**TABLE 1:** City of Weatherford demographic data

| Demographic data | Value |
| --- | --- |
| Population | 9,859 |
| Number of housing units | 3,991 |
| Percent with bachelor degree or higher | 37.1% |
| Unemployment rate | 3.0% |
| Median household income | US$26,908 |
| Percentage below poverty level | 11.8% |

As can be seen from Figure 2 and Tables 1 and 2, Weatherford provides an excellent location for an analysis of the impact of a wind farm. It is a relatively small city in a rural area and although its county has not experienced population loss, it is near the areas with an overall loss of population. Weatherford, Oklahoma is located in Custer County, approximately 80 miles west of Oklahoma City. Through innovations like using wind to power pumps for irrigation, the city has thrived as an agricultural community for the last several decades. Today, the majority of people make their living in educational or service-related fields followed by construction, manufacturing and agriculture (Tables 1 and 2, data obtained from the US Census Bureau). The addition of the wind farm has helped to add to the business climate and diversify the city's industrial image.

**TABLE 2:** City of Weatherford employment by sector

| Type of employment | Percentage |
|---|---|
| Management, professional occupations | 30.8 |
| Sales and office occupations | 30.8 |
| Service occupations | 16.2 |
| Construction, maintenance occupations | 10.6 |
| Production, transportation occupations | 10.2 |
| Farming, fishing, and forestry occupations | 1.4 |

This study analyzes the impact of the Weatherford Wind Energy Center. It is located in West Central Oklahoma, in Custer County, near and around the City of Weatherford. The developer is NextEra Energy and American Electric Power is the purchaser. Located on approximately 5,000 acres of land, the wind farm has 98 GE 1.5 MW turbines with a rated capacity of 147 MW of electricity—enough to power 44,000 homes. The wind regime is rated in the 'good' to 'excellent' range, according the wind resource class analysis, and consists of strong, consistent winds. The wind farm sits on low ridges that are higher in elevation than the city. These provide an additional increase in the winds as compared to the surrounding plains and, thus provide a suitable location for the wind farm. The location of this wind facility is highly visible to the public. Most of the 98 turbines can be seen from the state's major east–west Interstate (Interstate 40).

## 7.2 METHODS

The aim of this study is to use a multi-method approach to examine the impacts of the wind farm on Weatherford. This project consists of three components: economic modeling, surveys, and in-depth interviews. The specific methodological approaches are discussed below. Although there have been many studies examining economic impacts, as discussed above, and many others looking at public perceptions (e.g., see [10-12]), there are few if any studies that have provided a complete, holistic impact of a wind farm to a small community, and also, in addition, examined the social attitudes of the local populace as well as the key stakeholders and decision makers. In addition, there have been no previous studies in this area of the wind belt of the central US plains. This approach was taken since any one approach would not fully address the multidimensional aspect of the socioeconomic impacts. However, by combining approaches, a more detailed and robust picture of the impacts of the local wind farm can be determined.

### 7.2.1 ECONOMIC MODELING

The first component to assess the wind farm impact on the Weatherford area is an economic model analysis. The economic modeling is performed to determine direct economic impacts (e.g., increased local tax revenue) and indirect economic impacts (e.g., increased revenue from other industry sectors). For this study, the economic modeling was performed using a combination of the impact analysis and planning (IMPLAN) and job and economic development index (JEDI) input–output models. IMPLAN is an economic impact assessment modeling system, which can be used at many different geographic levels, from a state to county level. The initial intent of IMPLAN was to assist in land resource and land management. However, there are currently users representing a range of backgrounds from government to academia to private industry [13-16].

IMPLAN is an input–output model that relies on multipliers to quantify interactions between industries [17,18]. Each industry or service activity within the economy (e.g., agriculture, mining, manufacturing, and construction) is assigned to a specific sector (e.g., grain farming and fruit farming are assigned to agriculture; motor and generator construction are assigned to electrical equipment) within the economy. Input–output accounting describes commodity flow from the producer to the intermediate and final consumers. Total industry purchases including, for example, services, employment compensation, and imports, are equal to the good that is being produced [19]. This cycle of buying goods and services (indirect purchases) continues until leakage from the region stop the cycle. The additive features of these indirect and induced effects are compounded in the model through the Leontief Inverse Matrix [20]. The values in the Leontief Inverse Matrix represent the total direct and indirect requirements of any industry supplied by other industries within the region in order for that industry to be able to deliver US$1 worth of output to final demand [20]. Additional file 1 lists an example of some of the multipliers for Custer County.

For this study, the demographic and multiplier data was imported into IMPLAN to begin the model development. IMPLAN contains 528 economic sectors, and wind power is contained within the electric services sector. Wind energy makes up a very small percentage of this sector. To compensate for this, a sub-model of IMPLAN specific to the wind industry, the JEDI has been developed [21]. JEDI has been extensively used in wind energy impact modeling [22-25]. For example, Slattery et al. in 2011 [26] used JEDI to examine over 1,300 MW of wind farms in Texas. Results showed an increase of over 4,000 full time equivalent jobs and overall lifetime impact to the area of these wind farms of almost US$2 billion.

For this study, the county under analysis was selected within IMPLAN (Custer County in this case) and the model and multipliers were created, and this information was transferred into JEDI for further analysis. Several variables, such as project size, location, finance arrangements, and local economic factors influence construction and operating costs. The amount of local resources that are available can significantly impact the costs and economic impacts on a local region. Project specific data can be defined as a bill of goods; these components are considered critical in determining the number of jobs created. The bill of goods includes costs associated with actual construction of the facility, as well as annual operating and maintenance costs. To the extent possible, the model inputs were obtained for the specific project in Oklahoma; however, some of this data is proprietary, and thus in situations where data could not be obtained, best available estimates were used. For this research, the appropriate aggregation was created manually from the 528 sectors within IMPLAN. When the model aggregation was complete, the social accounts and the various multipliers were recomputed and entered in the JEDI model for final analysis.

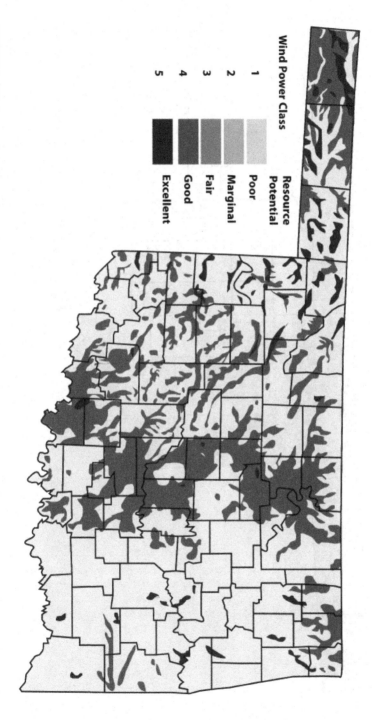

**FIGURE 1:** Oklahoma wind resource map.

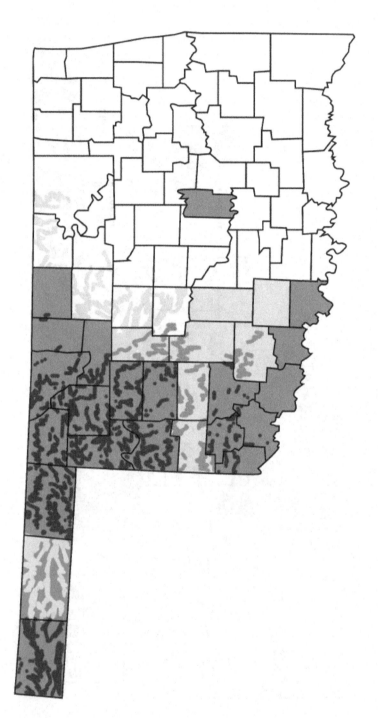

**LFIGURE 2:** Commercially viable wind and population loss counties.

## 7.2.2 SURVEY AND INTERVIEWS

IMPLAN and JEDI modeling provides useful information as part of an economic impact study. However, to try to present a more complete picture of the impact of the wind farm, qualitative methods were also used. First, direct interviews with some of the people responsible for the wind farms were undertaken in an attempt to add additional site-specific information. These interviews attempted to cover the significant aspects of wind farm development from public and elected officials as well as businesses within the community. Finally, a random survey of 108 adults was conducted in the community of Weatherford. This represents a cross-section of approximately 1% of the overall population. The goal of the survey was to gauge the level of knowledge regarding wind energy that people in the community had. This includes the wind industry in general and the Weatherford Wind Energy Center specifically. Respondents were asked a series of questions regarding their understanding of wind energy, and also a question regarding whether or not they thought the development hurt or helped property taxes. The surveys were handed out at randomly selected locations and times to make sure that the sample was as representative of the population as possible. Although there is no way, of course, to insure that there was no sample bias, we are confident that the bias has been minimized, and that the surveys do not have either a pro-or anti-wind bias.

## 7.3 RESULTS AND DISCUSSION

Results will be presented in order from most quantitative (e.g., the economic modeling) to most qualitative and descriptive (e.g., the interviews). This will provide not only a numerical assessment of the impact of the wind farm to the community but will also provide additional illustrative information.

## 7.3.1 ECONOMIC MODELING RESULTS

The economic analysis consists of efforts to characterize the impacts of the wind farm on Weatherford and focuses on the results of the combined

IMPLAN and JEDI modeling. Tables 3 and 4 represent the JEDI output. For this analysis, the construction cost in dollars per kilowatt (US$/kW) and annual operations and maintenance costs (US$/kW) are areas where the model shows sensitivity to changes [27]. A sensitivity analysis was conducted and the results suggested the best fit numbers to be used; those were the ones selected for the final analysis. In addition, the final parameters were determined in consultation with the wind farm operator. Results show that the 147 MW wind farm near Weatherford generated an estimated US$27 million in local spending and created 188 jobs during the construction phase. Once operational, the wind farm supports an estimated 13 jobs directly at the wind farm, including technicians and management. Furthermore, estimates show that US$1.7 million continues to be spent annually in the local economy, with over US$600,000 in additional property tax revenue and almost US$400,000 in direct land lease payments to landowners. The model estimates that the combined direct and induced impact annually is over US$25 million. The property tax is of particular importance, as this represents support for the local infrastructure (e.g., roads and schools) provided by the wind farm.

## 7.3.2 SURVEY RESULTS

Over 75% of the survey participants responded that they have some knowledge of wind energy. Fourteen percent of those surveyed felt they had a full understanding of wind energy. When asked if their knowledge had increased since the wind farm became operational, 79% of participants indicated that their knowledge had increased. The number that had little or no knowledge dropped down to just below 5%. This would suggest that because of the wind farm's high visibility, public knowledge of wind energy increased.

One area of concern for other wind farms has been whether or not the wind farms decreases property value. Opponents argue that property values drop when a wind farm is constructed in a community; however, there is no documented evidence that this is true [28-30]. For example, Sims et al. in 2008 report, 'no causal link was established between the presence of the wind farm and house price' Of course, the lack of a distance-price relationship only applies once a given marginal set-back distance has been maintained. Hoen et al. in 2011 state that 'neither the view of the wind facilities nor the distance of the home to those facilities is found to have a statistically significant effect on sales prices.' For Weatherford, 55% of the respondents felt that it had helped property taxes, indicating no evidence that people felt their property values had decreased since the wind farm development. In the next part of the survey, respondents were asked about their perception of the community, how well the State promotes renewable energy, and if they personally had benefited from the development of the wind farm. When participants were asked if they had a favorable opinion of wind energy, 85% responded yes, and less than 5% said they had a negative view. Nearly the same amount, 85%, felt the state should do more to promote wind and other forms of renewable energy. Less than 20% of participants felt that the state is adequately promoting renewable energy. One significant finding of this research was that when participants were asked if they felt their community was different than another community of similar size because of the wind farm, over 70% responded yes. This last point is interesting because of previous research that has been mixed about the impact of a wind farm on the local perceptions of an area [31,32]. Issues such as a negative viewpoint associated with the visual aesthetics do not seem to be present here, another indication of the overall widespread community support for the project.

When visiting Weatherford, it is clear that the local citizens exhibit noticeable civic pride in their community and that the wind farm has become a pivotal and productive facility now and for their community's future. However, in other locations, such as Europe, attitudes toward wind energy vary. This is often a 'not in my backyard' (NIMBY) perception. For example, visual evaluation is often mentioned as the most important factor for those opposed [33]. This type of NIMBYism is not evident in the study area in Oklahoma. In other areas, however, it is used by opponents of wind farms and often linked with wider environmental causes [34,35]. For example, Devine-Wright in 2005 conducted research examining the public's perception of wind farms, especially in areas where the NIMBY concept was the primary concern. This research examined the public's perception of the following: switching from conventional energy sources to renewable energy, wind turbines and people's negative perceptions of them, the physical proximity of wind turbines, the acceptance of wind farms over time, NIMBYism as an explanation of negative perceptions, and local involvement with these perceptions [36].

**TABLE 3:** JEDI output: wind plant - project data summary

| Data | |
|---|---|
| Project location | Custer County |
| Project size (MW) | 147 |
| Turbine size (kW) | 1500 |
| Number of Turbines | 98 |
| Construction cost (US$/kW) | 1,600 |
| Annual direct O&M cost (US$/kW) | 15.50 |
| Project construction cost (US$) | 235,200,000 |
| Local spending | 27,501,131 |
| Total annual operational expenses (US$) | 38,710,980 |
| Direct operating and maintenance costs | 2,278,500 |
| Local spending | 1,733,354 |
| Other annual costs | 36,432,480 |
| Local spending | 1,058,400 |
| Debt and equity payments | 0 |
| Property taxes | 666,400 |
| Land lease | 392,000 |

For the current study, one question that generated a wide range of answers was when participants were asked to provide their best estimate of the tax revenue Custer County received on a yearly basis from the wind farm. The amount the county can expect to receive each year will vary, but as Tables 3 and 4 show, this is estimated at over US$500,000. Participants had answers ranging from US$100 to US$1.75 million. The median value was approximately US$275,000. The large range in numbers suggests that the true economic impact of the wind farm is not as yet fully understood or realized. Fifty-three people or 49% did not provide an answer. People may have a perception and awareness of the wind farm, but are not well informed about the revenue that is coming into the city from the development. These numbers reflect that, and perhaps, the city can do better in highlighting the specific tax benefits from the wind farm to its citizens.

The final question asked participants to make a closing comment positive or negative on the wind farm. Below are some of the responses that were left by participants. These comments provide some additional qualitative context for how the residents of Weatherford view the wind farm. These comments illustrate a range of knowledge about the topic, but generally show the widespread support for the development.

**TABLE 4:** JEDI output: local economic impacts (dollar values in millions)

| Economic impacts | | | |
|---|---|---|---|
| During construction period | Jobs | Earnings (US$) | Output (US$) |
| Direct impacts | 4 | 9.2 | 26.9 |
| Indirect impacts | 84 | 1.8 | 5.7 |
| Induced impacts | 100 | 1.9 | 7.0 |
| Total impacts (direct, indirect, and induced) | 188 | 12.9 | 39.6 |
| During operating years (annual) | | | |
| Direct impacts | 19 | 11.8 | 17.5 |
| Indirect impacts | 61 | 1.3 | 4.4 |
| Induced impacts | 68 | 1.3 | 4.8 |
| Total impacts (direct, indirect, and induced) | 148 | 14.3 | |

## Comments from survey participants:

1. With the State as windy as it is it has to help.
2. It's crucial, it's beautiful. We need many more farms nationwide.
3. I am a little frightened by the giant wind mills.
4. The people that I have worked with have been very responsible people.
5. Who cares how it looks if it helps.
6. I think the wind farm is great!! It helps the people with turbines on their land and the economy of Weatherford.
7. I think it is wonderful for the environment.

8.  I think they are really neat to see and when you're coming back to Weatherford you know you are home.
9.  All Oklahoma communities should have wind generated energy, it wastes nothing and does no harm to the environment.
10. It has only helped economically, our community has yet to see a negative impact.
11. It's highly fantastic for our community.
12. I think it's a good thing for Weatherford.
13. Anything that saves energy is a good thing.


### 7.3.3 INTERVIEW ANALYSIS


The final piece of this case study consisted of a series of lengthy in-depth personal interviews. These interviews were with local politicians, public officials, and business owners. Over a dozen interviews were undertaken and the results here are indicative of the overall feelings of the stakeholders questioned. Not one local politician spoke out against the wind farm in our surveys and interviews. One key figure interviewed was the mayor of Weatherford. The mayor described the support of the community for the project. From the beginning, the mayor did not see any real opposition to the project, in fact in his words, he 'could count on one hand the number of people against it.' When the City was first approached, they examined it as they might any new development. The City benefited from knowing what had occurred in other communities with wind farms. The mayor was able to meet with city leaders from other locations and discuss how their communities had benefited. The mayor discussed some of the minor inconveniences such as torn up roads, but quickly added these were not significant in comparison to the long-term benefits. The City expects to fully benefit from new tourism associated with the wind farm and make Weatherford 'the Wind Capital of Oklahoma.' The mayor said that the state legislature had been cooperative in assisting Weatherford in any way it could and appreciated the balance of work that had been done at that level.

As anyone who goes to Western Oklahoma knows its wide open space and big sky produce nice and beautiful sunsets and horizons. The mayor was asked to put a value on that view and what may be lost when the turbines were installed, 'well, how do you put a price on such a thing?' the mayor asked. According to the mayor, to compensate the city and citizens of Weatherford, the wind farm developer agreed to pay the city US$25,000 a year for lost aesthetic beauty and community improvement projects. This is a huge benefit to the city and is not part of the city's normal budget. The city has used some of the money to install a security and surveillance system on the city government complex, a new city building, gym, and playground. Another interview was with the city's economic development manager and he could not have been more enthusiastic. He stated that all major sectors of the city's economy have benefited tremendously from the local wind farm development. Businesses, from hotels and apartment complexes to local restaurants, were all filled to capacity during construction.

In addition to city officials, a variety of local business owners were interviewed to examine how their business was impacted during the construction phase [37]. Local hotels reported that they were at capacity for three to four months at a time during the construction. For example, General Electric was a large client, sending representatives from Japan and Brazil to the area and renting blocks of rooms for a month at a time. This represented a significant impact for both the local Holiday Inn and the Comfort Inn. Other examples include Brundage Bone Concrete, Dolese Brothers Concrete, and Matt's Service Center. Sawatzky Construction benefited with nearly US$300,000 in revenue from the project including building a 5,000-ft$^2$ operation facility. The Southwest Fence Company supplied all fencing, cattle guards, and other security apparatus for the project. Matt's Service Center located just west of Weatherford along I-40 provided approximately 10,000 gal of diesel fuel and gasoline while also repairing damaged equipment during a two-month period during construction. This represented a total of US$100,000 in revenue during those 2 months. United Rentals saw increase in their revenue to US$70,000 through the rental of various pieces of heavy construction equipment; this equipment was rented for a five-month period.

## 7.4 CONCLUSIONS

When this research began, the overall goal was to assess the socioeconomic impact of wind farm development on a local community in Oklahoma through a multi-method approach. Communities that have similar characteristics as the one studied in this research should be able to gain information from this study and apply it.

The first effort was through the use of an economic input–output model. IMPLAN is an input–output modeling program that allows a user to input specific variables for economic analysis. The model results indicate that the county received a substantial economic impact during construction of the wind farm. The model-estimated impact shows the millions of dollars of both short-term and annual economic impact. Much of this money went to local construction companies in the community where the wind project was developed. The most important conclusion to be drawn from the economic modeling is that construction spending can be traced to two important variables. These variables are the size of project and the amount of goods and services that were purchased locally. The amount of goods and services that are purchased locally will ultimately have the greatest impact on a community during construction.

Custer County has already felt the impacts from wind development in Oklahoma. Many other counties across Western Oklahoma with similar wind resources have as well, and this represents the potential for millions of dollars in economic growth and new jobs for Oklahoma. In fact, Oklahoma is projected to continue to move up the list of top states with installed wind power by the end of the next decade. Using the most recent DoE projections, Oklahoma will reach as high as the second most important state in installed wind energy capacity by 2030 [38]. Thus, an industry that has a long history [39] will continue to play an increasing role in the development of western Oklahoma.

Economic modeling provides a quantitative description of the socio-economic impact. However, any research that is specific to an area or region also requires direct interviews with local officials and wind farm developers. The interviews that were conducted with community leaders and officials in Weatherford were very informative. There is no doubt that Weatherford has been positively impacted from the wind farm development. Interviews with local officials led to specific evidence of how the community had been impacted, including increases in tax revenue that have been used by the local school districts and other county entities that are essential for a healthy community. Other projects, such as community beautification, may prove difficult to accomplish if it were not for the wind facility operating in the community. Here is a quote from one community leader in western Oklahoma:

'After the wind farm was constructed outside of town, me and a co-worker had the notion to just get a couple of lawn chairs, a bottle of wine and just sit back and listen to the peace and silence that we have known our entire lives, interrupted by the brief, swoosh, and the enormous wind turbine blades cut through the air, and sit back with a smile on our faces and know that our grandchildren and their children have a more secure future because of the economic benefit of the wind turbines.'

This illustrates the impact not only in terms of economic numbers, but also in terms of the view of the community. One final illustration is that Weatherford has advertised itself as wind energy capital of Oklahoma where, to quote Rogers and Hammerstein, 'the wind comes sweeping down the plain.'

## REFERENCES

1. Righter R (1996) Wind energy in America: a history. University of Oklahoma, Norman.
2. DeCarolis JF, Keith DW (2006) The economics of large-scale wind power in a carbon constrained world. Energy Policy 34(4):395-410
3. Halperin A (2005) A shift in wind power? Business Week Online.
4. Schiermeier Q, Tollefson J, Scully T, Witze A, Morton O (2008) Electricity without carbon. Nature 454(7206):816-823 PubMed Abstract |
5. Barkley D (1995) The economics of change in rural America. Am J Agric Econ 77(5):1252-1258
6. Phimister E, Roberts D (2012) The role of ownership in determining the rural economic benefits of on-shore wind farms. J Agric Econ 63(2):331-360
7. Pedden M (2006) Analysis: economic impacts of wind applications in rural communities. National Renewable Energy Laboratory Subcontract report NREL/SR-500-39099. Report available at http://www.windpoweringamerica.gov/pdfs/wpa/econ_dev_casestudies_overview.pdf webcite, accessed 26 Jan 2013
8. National Wind Coordinating Committee (2004) A methodology for assessing the economic development impacts of wind power. National Wind Coordinating Committee, Washington D.C.
9. Brown JP, Pender J, Wiser R, Lantz E, Hoen B (2012) Ex post analysis of economic impacts from wind power development in U.S. counties. Energy Economics 34(6):1743-1754

10. Slattery MC, Johnson BL, Swofford JA, Pasqualetti MJ (2011) The predominance of economic development in the support for large-scale wind farms in the U.S. Great Plains. Renewable and Sustainable Energy Rev 16(6):3690-3701

11. Brannstrom C, Jepson W, Persons N (2011) Social perspectives on wind-power development in West Texas. Ann Assoc Am Geogr 101(4):839-851

12. Eltham DC, Harrison GP, Allen SJ (2008) Change in public attitudes towards a Cornish wind farm: implications for planning. Energy Policy 36(1):23-33

13. National Wind Coordinating Committee (2004) Economic development: impacts of wind, summary of case studies. National Wind Coordinating Committee, Washington D.C.

14. Costanti M (2004) Quantifying the economic development impacts of wind power in six rural Montana counties using NREL's JEDI model. National Renewable Energy Laboratory, NREL/SR-500-36414, Golden, CO.

15. Mulkey D, Hodges A (2004) Using IMPLAN to assess local economic impacts. University of Florida Institute of Food and Agricultural. Services, Gainesville.

16. Anderson D (1996) Economic effects of power marketing options in California's Central Valley: a 2005 impact analysis using IMPLAN. In: 1996 IMPLAN user's symposium. Battelle Pacific Northwest Labs, Minneapolis.

17. Becker S (2004) The impact of film production on the Montana economy & a proposed incentive for the film industry. Montana Department of Commerce. In: (2004) IMPLAN user's conference. Sheperdstown, West Virginia.

18. Braslau D, Johns RC (1998) Use of air transportation by business and industry in Minnesota. Air Transportation Res Rec 1998(1622):31-40

19. Lindall SA, Olson DC (2004) The IMPLAN input–output system. Minnesota IMPLAN Group Inc, Stillwater. PubMed Abstract |

20. Sonis M, Hewings GJD (1998) Temporal Leontief inverse. Macroeconomics Dynamics Camb Univ Press 2(1):89-114

21. Sinclair K, Milligan M, Goldberg M (2004) Job and Economic Development Impact (JEDI) Model: a user-friendly tool to calculate economic impacts from wind projects. In: National Renewable Energy Laboratory. 2004 Global Windpower Conference, Chicago, Illinois. 29–31 March 2004 NREL/CP-500-35953

22. Mongha N, Stafford ER, Hartman CL (2006) An analysis of the economic impact on Box Elder County, Utah, from the development of wind farm plants. U.S. Department of Energy, Golden.

23. Mongha N, Stafford ER, Hartman CL (2006) An analysis of the economic impact on Tooele County, Utah, from the development of wind power plants. Report for the U.S. Department of Energy, Energy Efficiency and Renewable Energy, August, No. DOE/GO-102006-2353.

24. Mongha N, Stafford ER, Hartman CL (2006) An analysis of the economic impact on Utah County, Utah, from the development of wind power plants. Report for the U.S. Department of Energy, Energy Efficiency and Renewable Energy, May, No. DE-FG48-05R810736.

25. Lantz E, Tegen S (2009) Economic development impacts of community wind projects: a review and empirical evaluation. Report for the National Renewable Energy Laboratory CP-500-45555.

26. Slattery M, Lantz E, Johnson BL (2011) State and local economic impacts from wind energy projects: Texas case study. Energy Policy 39(12):7930-7940

27. Williams SK, Acker T, Goldberg M, Greve M (2008) Estimating the economic benefits of wind energy projects using Monte Carlo simulation with economic input/output analysis. Wind Energy 11(4):397-414

28. Colwell PF (1990) Power lines and land value. J Real Estate Res 5:117-118

29. Sims S, Dent P, Oskrochi GR (2008) Modelling the impact of wind farms on house prices in the UK. Int J Strateg Prop Manag 12(4):251-269

30. Hoen B, Wiser R, Cappers P, Thayer M, Sethi G (2011) Wind energy facilities and residential properties: the effect of proximity and view on sales prices. J Real Estate 33(3):279-316

31. Jerpasen GB, Larsen KC (2011) Visual impact of wind farms on cultural heritage. A Norwegian case study, Environmental Impact Assessment Review 31(3):206-215

32. Clarke S (2009) Balancing environmental and cultural impact against the strategic need for wind power. Int J Herit Stud 15(2–3):175-191

33. Wolsink M (2007) Wind power implementation: the nature of public attitudes: equity and fairness instead of 'backyard motives. Renewable and Sustainable Energy Rev 11(6):1188-1207

34. Haggett C (2008) Over the sea and far away? A consideration of the planning, politics, and public perception of offshore wind farms. J Environ Policy and Planning 10(3):289-306

35. Jones CR, Eiser JR (2012) Understanding 'local' opposition to wind development in the UK: how big is a backyard? Energy Policy 38(6):3106-3117

36. Devine-Wright P (2005) Beyond NIMBYism: towards an integrated framework for understanding public perception of wind energy. Wind Energy 8(2):125-139

37. Cox C (2004) From snack bar to rebar: how project development boosted local businesses up and down the wind energy 'supply chain' in Lamar, Colorado. Report. U.S. Department of Energy, Washington D.C.

38. United States Department of Energy (2008) 20% wind by 2030. Report. United States Department of Energy, Washington D.C.

39. Kaldellis JK, Zafirakis D (2011) The wind energy (r)evolution: a short review of a long history. Renewable Energy 36(7):1887-1901

*There are several supplemental files that are not available in this version of the article. To view this additional information, please use the citation on the first page of this chapter.*

# CHAPTER 8

# Optimizing Wind Power Generation while Minimizing Wildlife Impacts in an Urban Area

GIL BOHRER, KUNPENG ZHU, ROBERT L. JONES, AND PETER S. CURTIS

## 8.1 INTRODUCTION

Many organizations have invested in clean energy or set targets for substituting a percentage of their power generation from renewable sources. Among renewable power sources, wind turbine energy is technologically mature and economically competitive. Typical economic considerations for wind farm locations are driven by long term wind statistics and expected energy output [1]–[4], interactions with other turbines [5], [6], and turbine height [7]. Environmental considerations, including collision risk with birds and bats, also affects the locations for wind farms [2], [3], [8]–[13].

The optimization of wind turbine location in urban areas, where small- and medium-sized wind turbines could be installed by families, organizations, municipalities, or other property owners is particularly challenging.

*Optimizing Wind Power Generation while Minimizing Wildlife Impacts in an Urban Area. © Bohrer G, Zhu K, Jones RL, and Curtis PS. PLoS ONE 8,2 (2013), doi:10.1371/journal.pone.0056036. Licensed under Creative Commons Attribution License, http://creativecommons.org/licenses/by/3.0/.*

In such limited-space urban applications, possible locations for such turbines are constrained by property boundaries and surrounding structures, particularly buildings and trees that affect wind flow in complex ways both locally and at high resolution [14]–[18]. The challenge often is compounded by difficulty in obtaining high resolution urban maps that include the height and shape of all wind obstructions [19], [20]. Moreover, it is of particular interest to avoid hazards to wildlife, such as bird and bat collisions [8], [21]–[28].

Given these constraints, optimizing turbine location in urban environments requires the incorporation of three disparate types of input data: a map of buildings, roads and other habitat types in the domain of interest, a map of power generation potential, and an assessment of environmental impact for each potential location [11], [26]. In limited-space applications, unlike for large-scale wind farms, the primary goal is not to extract the maximum amount of power from the region, but to find the best location for a limited number of turbines. Therefore, only locations with the best expected power output need be considered, unless other restrictions prevent placing turbines there. To achieve these ends, we propose to combine the application of exclusion zones—locations with the highest wildlife activity, but could include other environmental resources or ecosystem services, with high-resolution wind distribution data. While exclusion and buffer zones based on nature reserves or nesting sites have been proposed and applied before (e.g., [2], [3], [26]), they have never been proposed at high resolution in an urban settings. Using the exclusion-zone approach, the difference between the best power-generation potential across the full siting domain and the highest power generation potential at the remaining, unrestricted areas (after exclusion zones are applied) reflects the cost of avoiding negative environmental impact. As the threshold for acceptable environmental risk decreases, the size of the exclusion zone increases, and the maximal power generation potential in the remaining unrestricted area may either remain unaffected or decrease. The shape and rate of decrease of the maximal power curve with respect to wildlife risk provides a tool to evaluate tradeoffs between wind energy and environmental impact.

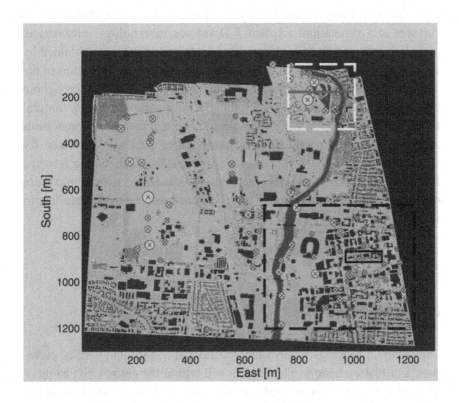

**FIGURE 1:** Cell type classification map of the OSU campus, simulation domains, and bird survey locations. Brown cells are buildings, red/orange/yellow are tall, medium and short trees respectively, dark blue is water surface, light blue is paved surface, light green is grass surface. The dark blue region at the outer edges of the map is an unclassified outer-edge buffer. The research wetland and central campus simulation domains are marked with yellow and black dashed frames, respectively. The bold black box marks the section illustrated in Appendix Fig. A.1. The red bar at the wetlands domain marks the location of the vertical cross-section illustrated in Fig. 3. Bird survey locations are marked with an x, and the radius of the circle around these locations is proportional to the total number of species observed at that location. X and Y axis represent eastward and southward coordinates in meters.

Here, we used the campus of the Ohio State University (OSU) in Columbus Ohio, USA, as an example of an urban application. The OSU campus, typical of many urban areas, includes buildings, open spaces, paved surfaces and vegetation. Explicit 3-D surface morphology information, including tree and building locations, height and shape, was obtained by combining publically available GIS data and airborne LIDAR scans of the campus. A high-resolution map of power generation potential throughout the campus was obtained by combining atmospheric large-eddy simulations with the long-term wind climatology. The spatial distribution of birds around campus during the summer season was used as a surrogate for wildlife activity to determine exclusion-zones. We demonstrate how the locations for which wind turbines should be considered to both maximize power generation and minimize collision hazard to wildlife were identified using the exclusion zone approach.

## 8.2 METHODS

### 8.2.1 PREPARING THE 3-D SURFACE MORPHOLOGY

Our goal was to assign a horizontal location, above-ground vertical elevation, and feature code to each $3\times3$ m cell within the entire OSU campus. This $3\times3$ m resolution was selected to match the resolution of the atmospheric model. Each cell was characterized as one object type: building, pavement, waterway, tree, or grass.

Data for ground elevation and object locations and shapes came from two sources: (1) Franklin County, Ohio, USA GIS data and (2) Ohio Statewide Imagery Program (OSIP) data. The Franklin county database was obtained from the Franklin County Auditor's office (www.franklincountyauditor.com) and includes data in ArcGIS and AutoCAD DXF formats. OSIP is a data product that was developed from processed airborne LIDAR scans. The OSIP data was downloaded from http://gis3.oit.ohio.gov/geodata/. The OSIP data included the following products: Ground Digital Terrain Model (DTM) source: $5000\times5000$ ft$^2$ ($1524\times1524$ m$^2$) tiles of the ground elevations at 1 ft$^2$ ($0.305\times0.305$ m$^2$) resolution; Original return source: $5000\times5000$ ft$^2$ irregularly dispersed at a distance of approximately 6.25 ft (1.904 m) on average; Background Images: 5000×5000 ft$^2$ (1524×1524 m$^2$) at 1 ft$^2$ (0.305×0.305 m$^2$) resolution, color geo-referenced MrSID images. These data sets were referenced using the Ohio State Plane South Nad83 Nav88 Survey feet coordinate system.

We used the GIS data to separate surface objects, such as buildings, roadways, parking lots, sidewalks, and waterways from other background data into a new data set. Polygons including roadway, parking lot, and sidewalk data were assigned properties of "pavement"; building layers formed a single feature type, "buildings"; and rivers, lakes, and ponds formed a single feature type, "waterways". We extracted tree information from the OSIP LIDAR data by comparing the original return elevations to the processed ground elevations. Cells where the LIDAR returns were at least 3 m higher than the ground elevation and that were not previously classified as buildings were classified as "trees". All remaining unclassified cells were classified as "grass". To insure that each GIS cell was only occupied once we created a hierarchy of feature dominance in the order: buildings, trees, pavement, water, and grass. Each cell in the resulting gridded 2D cell-type 3×3 m classification map was defined by its horizontal center and had an associated feature code (Fig. 1). Above-ground elevations were assigned to all features (S. Appendix Fig. A.1). Because trees are porous objects and it is difficult to specify their exact height, we separated tree cells into four bins: 5–10; 10–15; 15–20; and 20–25 meters.

## 8.2.2 PRODUCING A DETAILED WIND FIELD

We used an observation-based gridded field of winds from the North American Regional Reanalysis (NARR) [29] from 1979 until 2011. The dataset provides data at 3-hr snapshots at a spatial resolution of 0.3 degrees (roughly 32×32 km$^2$), and we used a model grid point (40.0748°N, 83.0896°W) located within the OSU campus.

We used the Regional Atmospheric Modeling System (RAMS)-Based Forest Large Eddy Simulations (RAFLES) [30], [31] to model the wind flow at OSU at high resolution. RAFLES is specifically designed to include the effects of vegetation and trees on wind flow and turbulence inside and above forest canopies. The RAFLES model solves the set of compressible

Navier–Stokes equations. The model is forced by a vertical profile of the mean horizontal wind, temperature and humidity, which are also used as the initial condition, and by surface heat and water vapor fluxes. We classified the NARR data for the OSU campus to 3 meteorologically-typical periods, characterized by a distinct combination of atmospheric conditions (Supporting information, Table S1), and conducted a set of RAFLES simulations for each of these typical periods. Each set was defined by a combination of vegetation density and surface heat flux and included 8 simulations with different wind directions. Simulations were at a resolution of 3×3 or 6×6 m horizontally and 3 m vertically. Buildings and trees were explicitly represented in the simulation domain (Supporting information, Fig. S1). Each model run simulated 2.5 hours for spinup and an additional 30 minutes in which results were analyzed. More details about the simulation and assumptions to reduce the number of simulations needed are provided in the supporting information Appendix S1. Due to the very high computational-time requirements of such a high-resolution model we did not simulate the entire campus environment but focused on two important areas—the central campus and the research wetlands (Fig. 1). Each model simulation provided information about the detailed wind field at one characteristic period and a specific wind direction. We used Monin-Obukhov surface similarity to scale the results of each simulation to different wind speeds within the same characteristic meteorological conditions. We combined all simulation results and scaled them to represent the entire long-term period. This was done by a weighted average of all wind fields. The weight for each windspeed-scale single simulation wind field was calculated based on the observed frequency (from the NARR dataset) of the meteorological conditions and wind direction and speed which that simulation result represented.

### 8.2.3 CALCULATION OF WIND-POWER POTENTIAL AT EACH LOCATION

A power curve was fitted into an empirical function, $f_o$, relating power output to specific mean wind speeds. $f_o$ is typically provided by turbine manufacturers. Here, we used $f_O$ of a relatively small, 1kW AWP-3.6 wind turbine (Aerofire Windpower, Lafayette, CO, USA. Power curve provided by www.solacity.com). The averaged potential wind power output of each pixel of the 3-D domain was calculated by integrating the power output, $W_{ij}$, of each wind speed bin:

$$W_{ij} = \sum_{j=1}^{n_j} P_j \sum_{i=1}^{4} P_i \sum_{n=1}^{42} f_O(S_n) \times (1 - \varphi) \times (1 - \rho) \times P_{n_{ij}}$$

(1)

where $S_n$ is the horizontal wind speed of the pixel, $P_j$ is the air density factor, $P_i$ is the probability of meteorological forcing in the historical dataset falling into each simulation category (Supporting information, Table S2), $P_{nij}$ is the probability of wind in that category blowing to one of the four direction bins, is the probability of the wind in that category and direction to be within a specific range (bin) of speed, and $\varphi$ is the turbulence factor. We used our simulation results to scale $\varphi$ at different locations. The mean relative turbulence level aloft (above 30 m) was assumed to scale as a turbulence factor of 5% and the relative turbulence level at the most turbulent places, such as in the wake of trees, scaled as a turbulence factor of 20%. All other values scaled linearly, between these two levels, based on the relative turbulence at each location.

### 8.2.4 BIRD SURVEYS

Bird observations were made at survey points every 200 m along transects running across OSU property (Fig. 1). We did not require a permit because no vertebrates were captured, handled or disturbed in this study. Birds were passively observed from public, urban areas where pedestrian traffic other than birdwatchers is common, and were not disturbed in any direct or indirect way by the observers. Observations were made for five minutes at each survey point by experienced avian biologists who recorded the species of bird seen or heard and the number of individuals of each species. We used all observations within 30 m (estimated using streets and buildings as visual references). This distance is approximately the size of a typical patch in the dense urban setting of this experiment. Each survey point was visited at least four times between June 14th and September 3rd, 2010, a period that included the peak activity season for local and summering birds in an effort to capture the maximal estimate for bird densities. Observations made during both the early and late morning. To characterize the survey environment, a 30 m-radius buffer circle was made around each survey point on the patch-type map. Then, the occurrence of each land-use patch type falling into the buffer circle was counted, yielding a relative density of each patch type for each bird-survey point.

**TABLE 1:** Empirical equations and line-fit statistics relating the observed number of individual birds or species with the density of the different patch types surrounding each location of the grid.

| Bird Variables | Line-Fit Coefficient by Patch Type | | | | | Line-Fit Const. | $R^2$ | P value |
|---|---|---|---|---|---|---|---|---|
| | Grass | Water | Pavement | Building | Tree | | | |
| All Species | 0.020 | 1.700 | | | | 3.650 | 0.24 | <0.001 |
| Native Species | 0.017 | 1.899 | | | | 2.789 | 0.19 | 0.001 |
| (All Individuals)$^{0.5}$ | | | −0.016 | | −0.022 | 5.223 | 0.08 | 0.026 |
| ln (Native Individuals) | | | −0.008 | −0.030 | | 2.537 | 0.18 | 0.011 |

*Only significant model results are shown.*

To extend the point observations of bird species richness and numbers of individuals to the entire campus area, we related the locations of bird observations to the specific environmental patch types around the survey point and fit an empirical model of bird numbers or species richness as a function of the environmental patch type. A stepwise forward multi-variant regression was used to find which of the patch types was significantly associated with larger numbers of individuals or species of birds (Table 1). Models were fitted for each of the four bird survey variables: (1) the mean number of total individuals; (2) mean number of native individuals; (3) mean number of total species; (4) mean number of native species. Native species often are the focus of conservation efforts while non-native species may be present in large numbers but generally are not considered of special conservation concern.

### 8.2.5 EXCLUSION ZONES AND UNRESTRICTED-DOMAIN POWER CALCULATION

The exclusion zone is an area that must be protected and therefore is excluded from consideration as a turbine site. Exclusion zones are determined as a prescribed fraction of the total domain with the highest wildlife activity density, or individual numbers or species richness endangered-species abundance. In this study we used the summertime activity of birds as an example for generating exclusion zone considerations. The mean power generation potential at the top (strongest power) 10% of the remaining un-excluded area provided a quantitative metric of the effect of the exclusion area, and we considered it as the power potential of the domain with a given exclusion level. We use the top 10% cutoff rather than the best point in the un-excluded domain because of the need for a large area from which placement could be selected. This is because other restrictions than bird presence or wind could prohibit construction of a turbine at a given site. We tested the effects of different exclusion thresholds of acceptable environmental risk by incrementally increasing the exclusion fraction.

## 8.3 RESULTS AND DISCUSSION

### 8.3.1 HIGH-RESOLUTION WIND DISTRIBUTION AND ITS IMPACT BY BUILDINGS AND VEGETATION

The 3-D map of power generation potential on selected areas on the OSU campus is shown in Figs. 2 & 3. Trees and buildings form a "wind shadow" effect, with weaker wind at lower elevations, near the canopy, relative to the same elevation over open areas (e.g., grass, water, and paved surfaces). However, due to the complex 3-D structure of the obstacles and their interactions, these wind shadow patterns do not follow an easy-to-define distribution downwind of each obstacle. Buildings tend to have a more pronounced downwind shadow than do trees as they block and deflect the wind more efficiently. This is even more pronounced in the neutral cases where surface heat fluxes do not play a role (data not shown).

It is interesting to note that in the Central Campus area (Fig. 2), the grass covered park-like area centered at 900 m east and 550 m south from the northwest corner of the domain in Fig. 2 has the highest wind power potential. The tall buildings of the medical center at the southwest region of the central campus block the wind and lead to a low wind power potential area. An interesting effect is generated by the two tall student dormitory towers at 450 and 650 m south and 100 m east (Fig. 2). The towers slow wind speed immediately above them, but funnel easterly and westerly winds to the grassy gap between the towers, and create a narrow corridor of relatively strong wind. The vertical profiles of power potential (Fig. 3) show that within 10 to 30 m above the ground, the power potential is almost double over open areas than over buildings or trees.

## 8.3.1.1 BIRD DISTRIBUTION

The relationship between bird survey variables and map patch types is shown in Table 1. On average we observed 12 species, 8 of which native and 28 individuals, 10 of which native at each observation location/time. The most abundant species included Canada goose and house sparrow. The presence of water had a strong positive effect on bird diversity and increased the predicted numbers of native and total bird species, but did not have a significant effect on the total number of individuals. Paved areas had a negative effect on individual numbers but not on species richness. Buildings had a negative effect on native species individual numbers, but not on total bird numbers. Surprisingly, trees had a negative effect on total numbers, but not on native species numbers. This is probably due to the large numbers of "city" birds, such as doves and sparrows that tend to be common in large numbers around buildings, and few other species, such as Canada geese that aggregate in large numbers on open grass lots.

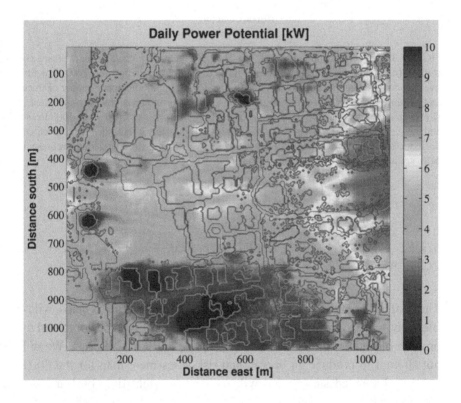

**FIGURE 2:** Daily power generation potential of the central campus area (black dashed frame in Fig. 1). Colormap shows the expected mean daily power potential [kW], darker lines mark the edges of land-use types and lighter lines mark the edges of buildings. X and Y axis represent eastward and southward coordinates in meters.

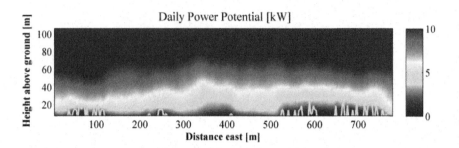

**FIGURE 3:** Vertical distribution of daily wind power generation potential over the research wetland (dashed frame in Fig. 1). This plot illustrates power generation at different heights over the cross-section illustrated as a line in Fig. 1. Lines mark the upper outline of trees and building. Stronger power potential can be found at lower elevations over open areas. However, as can be noticed by the difference between power over the trees at 100 m east and the shorter trees over 300 m east, the complex 3-D structure affects the height to which the reduction of power potential extends vertically over an obstacle. X and Y axis represent eastward and southward coordinates in meters.

## 8.3.1.2 MINIMIZING BIRD COLLISION RISK—THE EXCLUSION ZONE APPROACH

The goal of this study was to demonstrate how high resolution maps, wildlife activity density estimates and wind simulations could be combined to provide placement guidance for wind turbines in urban settings. We did not attempt to produce an operational risk-assessment map for the OSU campus. Extended survey that will include the night time bird and bat activity, and surveys during migration seasons will be needed in order to produce an accurate environmental assessment of wildlife activity and potential risk (see [32], [33] for the importance of OSU campus habitats during migration stopover).

We combined high-resolution data of different types for our urban study area using the exclusion zone approach. At the large (national, state) scale exclusion zones were proposed that restrict wind-power development in and near protected areas [2], [3], [11]. However, limited-domain, small-scale application represent a very different case. Distinct nature-conservation areas or major nesting colonies are typically not included anywhere in the domain, while a buffer distance of a few km from any bird habitat location will incorporate the entire domain. The best location for a wind turbine, while considering wildlife risk in a limited space of an urban or rural community, will have high wind power potential and low risk of bird mortality. Unfortunately, wildlife mortality rates due to wind turbines are known to be location and species specific [8], [22], [25]–[28]. Additionally, it is impossible to predict actual mortality rates before well-parameterized models of bird (or bat) movement that includes their location, height and activity [23], [34]–[36] exist for all species in the area. To avoid this complication, we assumed that risk would be proportional to bird activity density at a given location and used the bird activity during the summer season as an example.

The risk-density–proportionality assumption is commonly used in generating risk-assessment maps [37], [38]. The assumption that density of bird activity at the area around the zone of the potential wind-turbine blades is proportional to the collision risk is implicit in almost any flight model study that attempts to relate flight behavior to collision risk without actual mortality data [23], [34], [35]. Studies that used collision-mortality observations to evaluate this assumption provide contradictory evidence—some found support to the positive relationship between abundance and/or activity density and mortality [8], [39], [40], while others find poor relationships between bird density and mortality [41]. Some studies reported species specific effects, with a significant relationship between abundance and mortality in some species but not others at the same locations [42], [43]. This suggests that the species flight behavior and possible avoidance capabilities play an important role [40], [44], [45]. However, generating exact species-specific predictions for all species would not be feasible, particularly in a complex urban domain. We therefore accept the simple proportionally assumption as a practical solution.

We then considered the effect of varying the size of conservation exclusion zones on power output. At the "conservative" extreme case, we excluded all areas where bird numbers or species richness was higher than the 10th percentile, i.e. turbine placement was considered only in areas where bird numbers/richness were at the lowest 10%. We then relaxed this requirement in consecutive steps of 10% ending with a case that excluded only 10% of the area where bird numbers/richness was highest (Fig. 4).

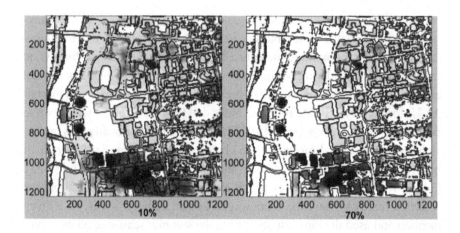

**FIGURE 4:** Effects of exclusion zones on expected power generation. The year-round power map at 30 m above ground (darker-high, lighter–low) overlaid on the map outline of the central campus area. Exclusion areas, 10%, and 70% of total domain with highest native bird density, are marked white.

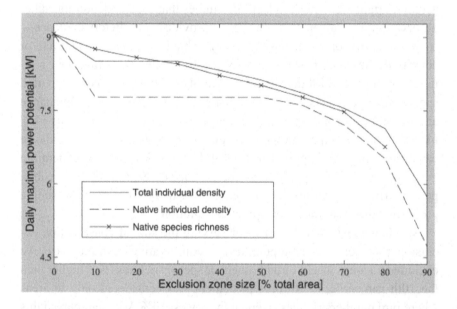

**FIGURE 5:** The effect of the relative area of exclusion zones on the maximal daily-mean power generation potential in the best remaining (un-excluded) suitable sites. Suitable sites are calculated as the 10% of the domain area with highest power potential.

As exclusion zones widen, the power generation potential of the top 10% of available locations decreases because more and more of the locations with high power generation potential are excluded. This decrease is non-linear when exclusion is based on number of individuals (Fig. 5). A sharp decrease (8% and 15% in total and native individual numbers, respectively) is caused by exclusion of the first 10% of the domain with the highest bird numbers or species richness. This is because the large grassy parks and the open water areas are the first to be excluded and also tend to have the highest power generation potential because of low drag and lack of obstructions. However, additional enlargement of the exclusion zone up to about 30–50% does not result in a large decrease of power generation potential at the best remaining non-excluded locations. These are typically found above low buildings, parking lots, and roads. This is not the case for species richness-based exclusion, however. In this case, a near linear decrease of power in the best remaining sites is driven by an increase in exclusion zone area.

Our bird observations were limited in space and time as is typical for direct observation animal surveys. Further developments in GPS-based tracking technology [46], particularly in miniaturization that will allow the tracking of small bird and bats is needed in order to incorporate the movement patterns and altitude of wildlife activity in future risk assessments and exclusion-zone considerations. Reduction of the tag prices will make it feasible to track many individuals in a single study and may facilitate wildlife tracking as a risk-assessment tool. Track annotation of birds and bats with turbulence and weather conditions [47]–[50] will allow turbine location decisions to incorporate the full tracks of birds and the behavioral rules according to which birds choose flight tracks and roost locations. These developments could yield more accurate movement models that could be applied at high enough resolution to be relevant to urban spaces.

## 8.4 CONCLUSIONS

We provide an example of a comprehensive data resource to support wind turbine placement decisions in a limited-space urban domain. Such decision support need is typical for university campuses, industrial complexes, farm cooperatives, or other entities that are considering adding wind turbines but want to minimize modification of the landscape in optimizing power generation. We used a large-eddy simulation (LES) in the context of fine resolution wind simulation with both vegetation canopy and buildings. The simulation result has many appealing features for future research. LES quantifies the explicit spatial effects of buildings and vegetation within the domain of a wind turbine. We showed that complex interactions between obstacles in different wind directions lead to a non-linear and complex patterns of wind speed at different heights above ground. LES results provide the information needed to find the location, as well as optimize the height, of the wind turbine. Alternatively, simpler foot-print models can provide information about the wind-shade of each obstruction given the wind speed and direction. While this will neglect the complex interactions between multiple obstructions, it will relax the need for a computationally expensive simulation and will allow resolving many more cases of different weather forcing.

Our study indicates a practical way of balancing the small-scale production of wind energy and minimizing wildlife collision risk. Combining the 3-D potential power generation map with the environmental-impact map leads to a location priority map that will provide planners with information on both the power output and environmental risk of a turbine application, with which they would optimize turbine location and height. In our example, areas supporting above-median bird numbers overlap with the areas where bird species richness is in the top 10–20 percentiles. Our analysis predicts that these areas could be excluded with only small consequences to the expected power generation potential at the best remaining locations.

## REFERENCES

1.  Chinnasamy TV, Haridasan TM (1991) Wind energy potential at Palkalainagar. Renewable Energy 1: 815–821. doi: 10.1016/0960-1481(91)90032-k
2.  El-Shimy M (2010) Optimal site matching of wind turbine generator: Case study of the Gulf of Suez region in Egypt. Renewable Energy 35: 1870–1878. doi: 10.1016/j.renene.2009.12.013

3.  van Haaren R, Fthenakis V (2011) GIS-based wind farm site selection using spatial multi-criteria analysis (SMCA): Evaluating the case for New York State. Renewable & Sustainable Energy Reviews 15: 3332–3340. doi: 10.1016/j.rser.2011.04.010

4.  Yu W, Benoit R, Girard C, Glazer A, Lemarquis D, et al. (2006) Wind energy simulation toolkit (WEST): A wind mapping system for use by the wind energy industry. Wind Engineering 30kmjn: 15–33. doi: 10.1260/030952406777641450

5.  Marmidis G, Lazarou S, Pyrgioti E (2008) Optimal placement of wind turbines in a wind park using Monte Carlo simulation. Renewable Energy 33: 1455–1460. doi: 10.1016/j.renene.2007.09.004

6.  Mosetti G, Poloni C, Diviacco B (1994) Optimization of wind turbine positioning in large windfarms by means of a genetic algorithm. Journal of Wind Engineering and Industrial Aerodynamics 51: 105–116. doi: 10.1016/0167-6105(94)90080-9

7.  Alam MM, Rehman S, Meyer JP, Al-Hadhrami LM (2011) Review of 600–2500 kW sized wind turbines and optimization of hub height for maximum wind energy yield realization. Renewable & Sustainable Energy Reviews 15: 3839–3849. doi: 10.1016/j.rser.2011.07.004

8.  Carrete M, Sanchez-Zapata JA, Benitez JR, Lobon M, Montoya F, et al. (2012) Mortality at wind-farms is positively related to large-scale distribution and aggregation in griffon vultures. Biological Conservation 145: 102–108. doi: 10.1016/j.biocon.2011.10.017

9.  Katsaprakakis DA (2012) A review of the environmental and human impacts from wind parks. A case study for the Prefecture of Lasithi, Crete. Renewable & Sustainable Energy Reviews 16: 2850–2863. doi: 10.1016/j.rser.2012.02.041

10. Kuvlesky WP, Brennan LA, Morrison ML, Boydston KK, Ballard BM, et al. (2007) Wind energy development and wildlife conservation: Challenges and opportunities. Journal of Wildlife Management 71: 2487–2498. doi: 10.2193/2007-248

11. Aydin NY, Kentel E, Duzgun S (2010) GIS-based environmental assessment of wind energy systems for spatial planning: A case study from Western Turkey. Renewable and Sustainable Energy Reviews 14: 364–373. doi: 10.1016/j.rser.2009.07.023

12. Leung DYC, Yang Y (2012) Wind energy development and its environmental impact: A review. Renewable and Sustainable Energy Reviews 16: 1031–1039. doi: 10.1016/j.rser.2011.09.024

13. Saidur R, Rahim NA, Islam MR, Solangi KH (2011) Environmental impact of wind energy. Renewable and Sustainable Energy Reviews 15: 2423–2430. doi: 10.1016/j.rser.2011.02.024

14. Celik A, Muneer T, Clarke P (2007) An investigation into micro wind energy systems for their utilization in urban areas and their life cycle assessment. Proceedings of the Institution of Mechanical Engineers, Part A: Journal of Power and Energy 221: 1107–1117. doi: 10.1243/09576509jpe452

15. Gao YF, Yao RM, Li BZ, Turkbeyler E, Luo Q, et al. (2012) Field studies on the effect of built forms on urban wind environments. Renewable Energy 46: 148–154. doi: 10.1016/j.renene.2012.03.005

16. Heath MA, Walshe JD, Watson SJ (2007) Estimating the potential yield of small building-mounted wind turbines. Wind Energy 10: 271–287. doi: 10.1002/we.222

17. Ledo L, Kosasih PB, Cooper P (2011) Roof mounting site analysis for micro-wind turbines. Renewable Energy 36: 1379–1391. doi: 10.1016/j.renene.2010.10.030

18. Mertens S (2002) Wind energy in urban areas: Concentrator effects for wind turbines close to buildings. Refocus 3: 22–24. doi: 10.1016/s1471-0846(02)80023-3

19. Cionco RM, Ellefsen R (1998) High resolution urban morphology data for urban wind flow modeling. Atmospheric Environment 32: 7–17. doi: 10.1016/s1352-2310(97)00274-4

20. Yu B, Liu H, Wu J, Hu Y, Zhang L (2010) Automated derivation of urban building density information using airborne LiDAR data and object-based method. Landscape and Urban Planning 98: 210–219. doi: 10.1016/j.landurbplan.2010.08.004

21. Arnett EB, Brown WK, Erickson WP, Fiedler JK, Hamilton BL, et al. (2008) Patterns of bat fatalities at wind energy facilities in North America. Journal of Wildlife Management 72: 61–78. doi: 10.2193/2007-221

22. de Lucas M, Ferrer M, Bechard MJ, Munoz AR (2012) Griffon vulture mortality at wind farms in southern Spain: Distribution of fatalities and active mitigation measures. Biological Conservation 147: 184–189. doi: 10.1016/j.biocon.2011.12.029

23. Katzner TE, Brandes D, Miller T, Lanzone M, Maisonneuve C, et al. (2012) Topography drives migratory flight altitude of golden eagles: implications for onshore wind energy development. Journal of Applied Ecology 49: 1178–1186. doi: 10.1111/j.1365-2664.2012.02185.x

24. Kikuchi R (2008) Adverse impacts of wind power generation on collision behaviour of birds and anti-predator behaviour of squirrels. Journal for Nature Conservation 16: 44–55. doi: 10.1016/j.jnc.2007.11.001

25. Kunz TH, Arnett EB, Cooper BM, Erickson WP, Larkin RP, et al. (2007) Assessing impacts of wind-energy development on nocturnally active birds and bats: A guidance document. Journal of Wildlife Management 71: 2449–2486. doi: 10.2193/2007-270

26. Masden EA, Fox AD, Furness RW, Bullman R, Haydon DT (2010) Cumulative impact assessments and bird/wind farm interactions: Developing a conceptual framework. Environmental Impact Assessment Review 30: 1–7. doi: 10.1016/j.eiar.2009.05.002

27. Noguera JC, Perez I, Minguez E (2010) Impact of terrestrial wind farms on diurnal raptors: developing a spatial vulnerability index and potential vulnerability maps. Ardeola 57: 41–53.

28. Osborn RG, Higgins KF, Usgaard RE, Dieter CD, Neiger RD (2000) Bird mortality associated with wind turbines at the Buffalo Ridge wind resource area, Minnesota. The American Midland Naturalist 143: 41–52. doi: 10.1674/0003-0031(2000)143[0041:bmawwt]2.0.co;2

29. Mesinger F, DiMego G, Kalnay E, Mitchell K, Shafran PC, et al. (2006) North American regional reanalysis. Bulletin of the American Meteorological Society 87: 343–360. doi: 10.1175/bams-87-3-343

30. Bohrer G, Katul GG, Walko RL, Avissar R (2009) Exploring the effects of microscale structural heterogeneity of forest canopies using large-eddy simulations. Boundary-Layer Meteorology 132: 351–382. doi: 10.1007/s10546-009-9404-4

31. Bohrer G, Nathan R, Katul GG, Walko RL, Avissar R (2008) Effects of canopy heterogeneity, seed abscission, and inertia on wind-driven dispersal kernels of tree seeds. Journal of Ecology 96: 569–580. doi: 10.1111/j.1365-2745.2008.01368.x

32. Rodewald PG, Matthews SN (2005) Landbird use of riparian and upland forest stopover habitats in an urban landscape. Condor 107: 259–268. doi: 10.1650/7810

33. Pennington DN, Hansel J, Blair RB (2008) The conservation value of urban riparian areas for landbirds during spring migration: Land cover, scale, and vegetation effects. Biological Conservation 141: 1235–1248. doi: 10.1016/j.biocon.2008.02.021

34. Eichhorn M, Johst K, Seppelt R, Drechsler M (2012) Model-based estimation of collision risks of predatory birds with wind turbines. Ecology and Society 17: ART. 1. doi: 10.5751/es-04594-170201

35. Masden EA, Reeve R, Desholm M, Fox AD, Furness RW, et al. (2012) Assessing the impact of marine wind farms on birds through movement modelling. Journal of the Royal Society Interface 9: 2120–2130. doi: 10.1098/rsif.2012.0121

36. Baisner AJ, Andersen JL, Findsen A, Granath SWY, Madsen KO, et al. (2010) Minimizing collision risk between migrating raptors and marine wind farms: development of a spatial planning tool. Environmental Management 46: 801–808. doi: 10.1007/s00267-010-9541-z

37. Bright J, Lanyston R, Bullman R, Evans R, Gardner S, et al. (2008) Map of bird sensitivities to wind farms in Scotland: A tool to aid planning and conservation. Biological Conservation 141: 2342–2356. doi: 10.1016/j.biocon.2008.06.029

38. Drewitt AL, Langston RHW (2006) Assessing the impacts of wind farms on birds. Ibis 148: 29–42. doi: 10.1111/j.1474-919x.2006.00516.x

39. Everaert J, Stienen EWM (2007) Impact of wind turbines on birds in Zeebrugge (Belgium). Biodiversity and Conservation 16: 3345–3359. doi: 10.1007/s10531-006-9082-1

40. Barrios L, Rodriguez A (2004) Behavioural and environmental correlates of soaring-bird mortality at on-shore wind turbines. Journal of Applied Ecology 41: 72–81. doi: 10.1111/j.1365-2664.2004.00876.x

41. de Lucas M, Janss GFE, Whitfield DP, Ferrer M (2008) Collision fatality of raptors in wind farms does not depend on raptor abundance. Journal of Applied Ecology 45: 1695–1703. doi: 10.1111/j.1365-2664.2008.01549.x

42. Ferrer M, de Lucas M, Janss GFE, Casado E, Muñoz AR, et al. (2012) Weak relationship between risk assessment studies and recorded mortality in wind farms. Journal of Applied Ecology 49: 38–46. doi: 10.1111/j.1365-2664.2011.02054.x

43. Smallwood KS, Thelander C (2008) Bird mortality in the Altamont Pass Wind Resource Area, California. Journal of Wildlife Management 72: 215–223. doi: 10.2193/2007-032

44. Smallwood KS, Rugge L, Morrison ML (2009) Influence of behavior on bird mortality in wind energy developments. Journal of Wildlife Management 73: 1082–1098. doi: 10.2193/2008-555

45. Garvin JC, Jennelle CS, Drake D, Grodsky SM (2011) Response of raptors to a windfarm. Journal of Applied Ecology 48: 199–209. doi: 10.1111/j.1365-2664.2010.01912.x

46. Rinne J, Riutta T, Pihlatie M, Aurela M, Haapanala S, et al. (2007) Annual cycle of methane emission from a boreal fen measured by the eddy covariance technique. Tellus Series B-Chemical and Physical Meteorology 59: 449–457. doi: 10.3402/tellusb.v59i3.17009

47. Dodge S, Bohrer G, Weinzierl R (2012) MoveBank track annotation project: linking animal movement data with the environment to discover the impact of environmental change in animal migration. In: Janowicz K, Keßler C, Kauppinen T, Kolas D, Scheider S, editors. Workshop on GIScience in the Big Data Age In conjunction with the seventh International Conference on Geographic Information Science 2012 (GIScience 2012). Columbus, OH: GIScience 2012. pp. 35–41.

48. Bohrer G, Brandes D, Mandel JT, Bildstein KL, Miller TA, et al. (2012) Estimating updraft velocity components over large spatial scales: contrasting migration strategies of golden eagles and turkey vultures. Ecology Letters 15: 96–103. doi: 10.1111/j.1461-0248.2011.01713.x

49. Kranstauber B, Cameron A, Weinzierl R, Fountain T, Tilak S, et al. (2011) The Movebank data model for animal tracking. Environmental Modelling & Software 26: 834–835. doi: 10.1016/j.envsoft.2010.12.005

50. Mandel JT, Bohrer G, Winkler DW, Barber DR, Houston CS, et al. (2011) Migration path annotation: cross-continental study of migration-flight response to environmental conditions. Ecological Applications 21: 2258–2268. doi: 10.1890/10-1651.1

*There are several supplemental files that are not available in this version of the article. To view this additional information, please use the citation on the first page of this chapter.*

# Land Cover and Topography Affect the Land Transformation Caused by Wind Facilities

JAY E. DIFFENDORFER AND ROGER W. COMPTON

## 9.1 INTRODUCTION

Wind energy is one of the fastest growing segments of the electricity market in many nations and this trend will likely continue as countries strive to reduce greenhouse gas emissions while meeting growing energy demands [1]. Research continues to predict the environmental consequences of new wind generation [2]–[4], compare impacts of different energy production technologies [5], consider methods for optimal placement of wind facilities [6], [7] and study the role wind should play in future energy strategies [8]–[11]. All of these activities require fundamental information that relates wind installations to the energy they produce and their impacts (both positive and negative) on economic, cultural, and environmental systems.

One impact of wind facilities is land transformation caused by building and maintaining the facility. These changes to land surface, including roads, can create management concerns for species affected by habitat loss and fragmentation [12]. We follow Fthenakis and Kim [5] use of "land

*Land Cover and Topography Affect the Land Transformation Caused by Wind Facilities. Diffendorfer JE and Compton RW. PLoS ONE 9,2 (2014), doi:10.1371/journal.pone.0088914. The work is made available under the Creative Commons CC0 public domain dedication, http://creativecommons.org/publicdomain/zero/1.0/.*

transformation" and use ha/MW to define it following Denholm et al. [13]. Thus, land transformation describes the amount of land transformed to produce one MW of capacity.

**TABLE 1:** Reported values of land transformation associated with wind development, the original value transformed into ha/MW, and the predicted amount of land transformation required to meet 251GW of land based wind energy—a stated goal of the Department of Energy (DOE) [9].

| Studies | Reported value | ha/MW | km² for 251GW |
|---|---|---|---|
| Roads and infrastructure only | | | |
| Denholm et al. [13] (low) | 0.06 ha/MW | 0.06 | 151 |
| BLM [23] (low) | 0.4 ha/ 1.5 MW turbine | 0.27 | 678 |
| Denholm et al. [13] (mean) | 0.3 ha/MW | 0.3 | 753 |
| DOE [9] (low) | 2% of 5MW/km² | 0.4 | 1,004 |
| BLM [23] (high) | 1.2 ha/1.5 MW turbine | 0.8 | 2,008 |
| DOE[9](high) | 5% of 5MW/km² | 1 | 2,510 |
| Denholm et al. [13] (high) | 2.4 ha/MW | 2.4 | 6,024 |
| Entire project area | | | |
| Denholm et al. [13] (low) | 4.76 ha/MW | 4.76 | 11,948 |
| DOE [9] | 5 MW/km² | 20 | 50,200 |
| Denholm et al. [13] (mean) | 34.5 ha/MW | 34.5 | 86,595 |
| Elliot [25] (high) | 2.65 MW/km² | 37.7 | 94,717 |
| Mackay [26] | 2W/m² | 50 | 125,500 |
| Elliot (low) | 1.03 MW/km² | 97.1 | 243,689 |
| Pimentel et al. [3] | 1 billion kwh/yr/13700ha | 120.1 | 301,683 |
| Denholm et al. [13](high) | 135 ha/MW | 135 | 338,850 |

*The reported values are organized based on the lands directly transformed by roads and infrastructure development and the entire project area. Some studies reported ranges of values and means, labeled as " low," "mean," and "high."*

To date, studies of wind power utilize estimates of land transformation for GIS-based modeling [4] or basic calculations of the land area required to generate energy using wind [9], [14]. However, published estimates vary by more than 1000 times (0.06 to 135 ha/MW, Table 1) with the estimated amount of area needed to meet the United States Department of Energy's land-based wind production goal of 251GW [9] ranging from ~151 to 338,000 km². These levels of uncertainty make analyses using point estimates subject to scrutiny and comparisons to other forms of energy extremely difficult.

Why do these estimates vary so much? Different methodologies explain some of the variation in reported land transformation. For example, methods including just the direct land transformation of a facility produce much lower estimates than methods that define the area as a polygon around the facility or the land leased by the facility (Table 1). However, even within each of these general approaches, estimates of land transformation vary substantially. We note that none of the studies in Table 1 actually measured surface disturbance from working wind facilities using modern GIS-based approaches. Instead, calculations often assumed a relationship between turbines and surface disturbance, used planning documents to estimate disturbance, or made assumptions about turbine spacing and land requirements.

Another explanation is that factors such as turbine size, topography, and pre-development land cover affect the land transformation caused by wind facilities. All of these variables could influence the distance between turbines, the amount of new roads, and the land cleared to install and operate a facility. For example, Denholm et al. [13], in perhaps the most thorough study to date, estimated land transformation from 172 environmental impact reports of wind facilities and reported average values by: turbine string configuration, land cover, and the total nameplate capacity of the facility. The reported mean values suggested land transformation of wind facilities were influenced by geographic variables and string configuration though no formal statistical analyses were performed.

Table 1 indicates the land transformation associated with wind facilities is perhaps poorly measured and not well understood. Understanding factors affecting land transformation will make future forecasts of the spatial impacts of wind power more accurate. Furthermore, if geographic features such as topography and land cover affect the land transformation caused by wind facilities, decision makers can use this knowledge to plan facilities that maximize energy production while minimizing surface impacts.

We report results from a geospatial analysis of 39 wind facilities we fully digitized using high resolution photo-imagery. We selected the facilities and designed our analyses to elucidate the effects of turbine size, turbine string configuration, topography, and land cover on land transformation. The results indicate the high levels of variation in land transformation across facilities can be explained, in part, by geographic variables.

## 9.2 MATERIALS AND METHODS

We digitized wind facilities in the United States selected to span gradients in turbine size, land cover, topography, and string configuration. We used turbine size information for each facility from the Energy Information Agency (EIA) to categorize facilities into four turbine size classes (<1.5, 1.5 to <2.0, 2.0 to <2.5, and >2.5 MW). When we began the work, turbine location data were not publicly available, as it is now, from the Federal Aviation Agency. We searched for facilities based on turbine size and the limited locational data in the EIA database, and other sources (Google Earth, the websites of wind energy companies, and county-level information). When we found a location, we digitized turbine locations using 1-m resolution USDA/NRCS Digital Orthophoto Quad Imagery (DOQ) county mosaics (http://datagateway.nrcs.usda.gov/) and then verified our turbine counts with those in the EIA database. We also verified facilities by matching photographs from company websites with aerial imagery, measuring turbine blades to verify make and model of turbines, matching this to facility information, and contacting county agencies for siting information when necessary.

Facilities were categorized using visual interpretation of aerial imagery. Land cover included categories describing both the general vegetation type and human use of the land. The categories were forest, shrub, grassland, hay, and tilled. Hay locations were mowed but not tilled, whereas grasslands were naturally occurring grasslands typically used for grazing. Topography included simple categories of flat, hills, ridgelines, and mesas. Flat locations had almost no topographic relief, whereas areas with turbines across numerous hills were categorized as hills. In many cases turbines were placed to follow the edge of a ridge or cliff. We categorized these as ridgeline. Some facilities were placed on mesas, an elevated area of land with a flat top and steep cliffs. When the turbines only followed the cliff edge, we considered these ridgelines, but when turbines were along both the cliff edge and across the top of the mesa, we considered these mesas. See File S1 for examples of facility classifications. We followed Denholm et al. [13] and categorized the configuration of turbines at a facility as single string, parallel strings, multiple strings, or clustered. Multiple string facilities included more than one string of turbines but these were not in parallel lines whereas clustered facilities had turbines scattered across an area with no obvious strings.

**TABLE 2:** Sample sizes associated with each categorical variable used in the analyses.

| Variable | Category within variable | | | | |
|---|---|---|---|---|---|
| Land use/cover | Forest | Grassland | Hay | Shrub | Tilled |
| | 7 | 5 | 8 | 9 | 10 |
| Topography | Flat | Hills | Ridgelines | Mesa | |
| | 15 | 4 | 13 | 7 | |
| Turbine Size | < 1.5MW | 1.5–< 2.0 MW | 2–< 2.5 MW | > 2.5 MW | |
| | 9 | 12 | 11 | 7 | |
| String Configuration | Clustered | Multiple | Parallel | Single | |
| | 6 | 15 | 6 | 12 | |

*In all cases, total sample is 39.*

Finding facilities across all combinations of explanatory variables was impossible. For example, facilities with 3MW turbines in both flat and forested sites simply did not exist. As such, we could not design a fully orthogonal study. Instead, we balanced the objective to sample across the categories described above with the difficulty of finding and verifying facilities and our budget. In addition, we avoided using multiple phases of development or closely spaced facilities as replicates. This process reduced issues related to pseudoreplication, but also limited our sample sizes in some cases. Though RWC digitized nearly 17,000 turbines, we ultimately could identify and use 39 facilities. Table 2 describes sample sizes by categories used in the analyses.

## 9.2.1 DIGITIZING PROCEDURES

We focused on estimating new land transformation caused by the development of a wind facility, not all of the infrastructure necessary to install, run, and maintain a facility. We did this for two reasons. First, we could not confidently understand the complete set of infrastructure required to support a facility from aerial photography, particularly if some infrastructure did not change during the installation of the facility. Second, we wanted to understand and quantify if placing facilities in areas with preexisting infrastructure, such as a robust road network, affected the amount of new infrastructure required.

Knowing the installation date of each facility, we compared ortho-imagery prior to, and after installation to determine the land transformation. We digitized any land transformation caused by the facility installation including new or widened roads, gravel pits, staging and storage areas, communication or meteorological towers, turbine pads and string roads, buried cables, disturbance to the original landscape caused by landslides, road berms, or erosion control treatments, roads and vegetation clearing associated with aerial transmission, and electricity substations (File S1).

We approached land transformation mapping conservatively. For example, in cases where vegetation recovered rapidly, or in agricultural areas where buried cables did not affect future farming activities, we did not include these short-lived disturbances. We did not include the original disturbance associated with the facility if vegetation had recovered (based on aerial imagery) within five years or if the original land cover (farming) was still viable (File S1, Slide 1). In general, this only happened in tilled locations or in moist grasslands where grasses recovered rapidly. We also treated roads conservatively, by subtracting preexisting roads from new or widened roads when the two overlapped. Finally, we only digitized surface disturbance if it could be linked specifically to the wind facility (File S1, Slide 2). For example, some facilities shared an electricity substation with other facilities, or road networks linked different phases of facility development. When questions arose regarding what facility caused a specific transformation, we always erred conservatively and did not include the surface change in our digitizing.

## 9.2.2 STATISTICAL ANALYSIS

The facilities we digitized were highly variable in their structure. Some had extremely long access roads, buried cables, or transmission lines that generally connected the facility to existing roads or power lines. These idiosyncratic connecting infrastructures made large contributions to a facilities land transformation. To address this issue, we analyzed facilities at three spatial scales: "strings," "sites," and entire "facilities." "Strings" represented the smallest spatial unit and consisted of turbine pads and the roads and buried cables connecting them. Strings allowed us to analyze relationships between turbine size and land transformation without the addition of jump roads between strings, gravel pits, and other forms of surface disturbance. "Sites" included the strings, the roads or electrical connections between strings, and other infrastructure associated with the strings. Long access roads or connections to the larger power grid were not included in sites. Finally, "facilities" included all digitized land transformation.

We estimated the total area of land transformation and the area attributed to roads for each facility at each spatial scale. To estimate land transformation, we divided the area of land we mapped as transformed (ha) by the facilities nameplate capacity (MW). We also estimated turbine spacing by calculating the average nearest-neighbor distance between turbine locations at a facility.

We use generalized linear models to analyze the effects of turbine size, land cover, topography, and string configuration on land transformation at each spatial scale. Unlike land transformation, turbine locations were fixed and did not vary with spatial scale, so analyses of the mean nearest-neighbor distance between turbines was only done once.

Generalized linear models are a class of statistical models that share a number of properties, such as linearity and computational approaches for estimating parameters [15]. Our generalized linear models were structured so that either land transformation or nearest-neighbor distance between turbines were the dependent variables, and combinations of turbine size, land cover, topography, and string configuration were included as categorical explanatory variables. The models included a Gaussian error distribution and an identity link function. Sample size was 39 facilities for all models whereas the number of parameters varied depending on the combination of explanatory variables included.

Because multiple explanatory variables may drive the land transformation of wind facilities simultaneously, our analyses were based on the approach of multiple working hypotheses [16], [17]. Model selection based on information theory is well suited for these types of analyses [18], and we followed well-described approaches [19]. We performed model selection by developing a list of candidate models that included both individual and additive combinations of explanatory variables then ranked models using Akaike's Information Criterion adjusted for finite sample sizes (AICc). Model selection was based on the differences in AICc across models where models within 2 AICc were considered to have substantial support. We also used AICc weights, which range from 0 to 1, and represent the weight of evidence that a model is the best model given the data and the set of candidate models. To understand the relative importance of explanatory variables we summed the AICc weights across all the models where the variable occurred. Larger values of the summed AICc weights indicate which explanatory variables most affected land transformation or turbine spacing [20]. We used model averaging to estimate effect sizes and report means and estimates of precision based on unconditional standard errors [19], [21]. All analyses were done in R [22].

## 9.3 RESULTS

Similar to previously reported values, land transformation varied considerably across facilities. The range declined from the facility scale (0.11–4.3 ha/MW) to the site (0.11–1.5 ha/MW) and string (0.11–1.3 ha/MW) scales illustrating how road and transmission lines that connect facilities to transportation networks and the electric grid can increase land transformation.

### 9.3.1 LAND TRANSFORMATION

Model selection indicated land cover and topography, but not turbine configuration or size, were generally associated with the amount of land transformation across facilities. Land cover or topography commonly occurred in the best-supported models and had the highest parameter AICc weights (Tables S1–S3). An exception was at the facility scale, where topography had low AICc weights and was not in any of the most supported models, whereas turbine size had higher AICc weights and was in the best-supported models (Table S3). This result was caused by the long access road and transmission line at Kibby Mountain, which also had large, 3MW turbines. At the facility scale, Kibby Mountain, with an extremely large land transformation, affected model rankings, enhancing the role of turbine size. Given this unique situation, we doubt this result is generalizable. At the other spatial scales (Tables S1 and S2) turbine size was not found in the best-supported models, and its variable weights were small, indicating little to no effects on land transformation.

At all spatial scales, facilities on tilled landscapes had lower levels of land transformation than forests and shrublands (Fig. 1). Facilities on mesas had higher land transformation values than facilities in flat topographies at the string and site scales (Fig. 2). Topography did not affect land transformation at the scale of entire facilities.

Mean land transformation at the facility ($0.93\pm0.12$ ha/MW, mean+SE) and the site scales ($0.72\pm0.06$ ha/MW) were two to three times higher than the mean value (0.3 ha/MW) derived from a review of environmental impact reports (Table 1, see Denholm et al. [13] for road and infrastructure only). At all scales, the surface of new roads (not berms or other disturbances associated with roads) accounted for 38–41% of a facilities' land transformation (facility scale: $41\%\pm0.04$, site scale: $38\%\pm0.04$, string scale: $41\%\pm0.04$, mean$\pm$SE).

## 9.3.2 TURBINE SPACING

Land cover and turbine size affected turbine spacing. Tilled locations ($357.7\pm48.2$ m, mean $\pm95\%$ confidence interval had higher turbine spacing than other types of land cover (Forest = $226.7\pm60.3$, Grassland = $219.8\pm70.8$, Hay = $211.7\pm57.6$, Shrub = $211.32\pm53.7$). In addition, larger turbines had higher distances between them (Fig. 3, Table S4).

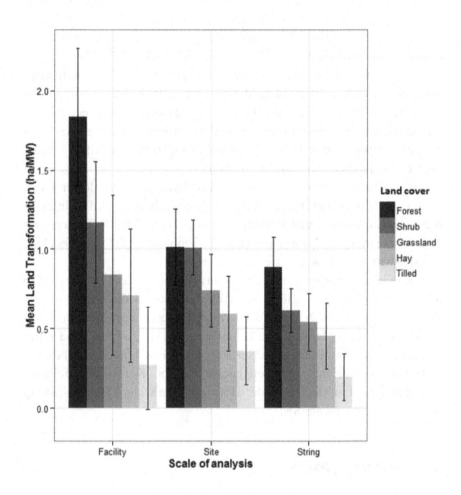

**FIGURE 1:** Mean (±95% Confidence Interval) of the land transformation associated with wind facilities in different land use and cover ("Land cover") categories at 3 spatial scales of analysis

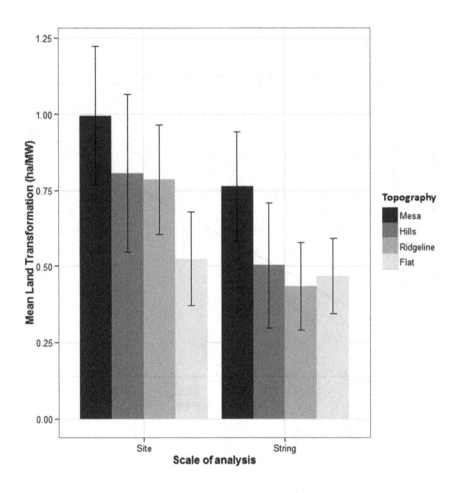

**FIGURE 2:** Mean (±95% Confidence Interval) of the land transformation associated with wind facilities in different topographic categories at 2 spatial scales of analysis.

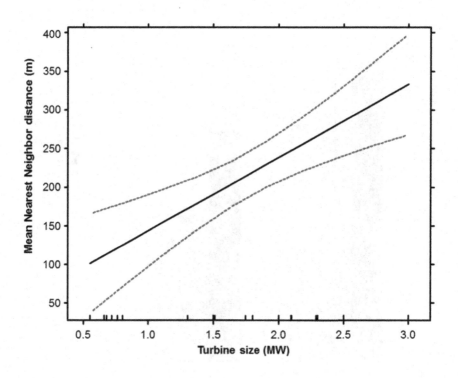

**FIGURE 3:** The linear relationship (black line) and 95% Confidence Interval (red lines) between the mean nearest neighbor distance among turbines at a facility and the size of the turbines in MW of capacity.

## 9.4 DISCUSSION

The results indicate the levels of land transformation associated with wind energy vary substantially across facilities. Furthermore, geographic variables are more likely to drive levels of land transformation than turbine-specific variables. Where a facility was built affected the amount of land transformed more so than the size of the turbines or how they were configured. Furthermore, our statistical models produced estimates of effect sizes for those variables driving land transformation, allowing predictions of the land transformation associated with wind energy.

Wind facilities can change through time and space as turbines are added or removed. Our study represents a single snapshot of land transformation from industrial scale wind energy. The long connecting roads and transmission lines at some facilities had large impacts on their land transformation. For example, Kibby Mountain, with the highest land transformation at the facility scale (4.3 ha/MW) had a ~47 km transmission line and a 14 km road (File S1, Slide 3). If these were installed to support multiple phases of development, then future estimates of land transformation at Kibby Mountain may decrease. We performed our analyses at three spatial scales in part because connecting infrastructure increased land transformation, yet the use of long roads and power lines appeared sporadically and with no apparent patterns across facilities. Despite this, land cover affected land transformation at all three spatial scales, whereas topography affected land transformation at the site and string scales.

At least two processes may explain why tilled landscapes had lower levels of land transformation. First, tilled landscapes have a large road network capable of supporting heavy agricultural machinery. Facilities in tilled landscapes use these existing roads, typically building a small road to connect an individual turbine to a preexisting road. Second, land was replanted in agricultural fields where cables were buried, as were road berms or small scrapes near the turbine pad (File S1, Slide 1). Other land cover types did not have the robust road networks, and surface disturbance associated with facility installation was not rapidly re-used.

Land transformation in forests was generally higher than other land cover types. At the facility scale, this was likely driven by the high land transformation associated with Kibby Mountain. However, even at the site and string scales, land transformation at facilities in forests was generally higher than other land cover types. Larger clearings around turbines, roads, and intersections caused by tree and canopy removal may drive this pattern (File S1, Slides 3 and 4). For example, trees must be cleared in wide arcs around intersection corners to make room for turbine blades on flatbed trucks. In addition, trees were cleared around pads and roads at what appeared to be distances long enough to avoid hazards associated with tree falls.

We expected topography to affect land transformation by requiring longer and more sinuous roads in areas with higher levels of relief. At the site scale, flat geographies had the lowest average land transformation. However, mesas were generally flat yet facilities on mesas had the highest average land transformation. We are uncertain of the underlying mechanism, but placing turbines along ridge edges in combination with additional turbines across tops of mesas, or other flat areas may require more complex road networks (File S1, Slide 5).

One might expect greater distances between turbines would require longer roads and hence more land transformation. However, tilled landscapes had the largest distances between turbines yet the smallest land transformation. Unlike larger tracts of land with parallel lines of turbines, turbines were frequently more scattered in agricultural areas. We suspect lease agreements with individual landowners and zoning laws such as road and residence setback requirements, may result in wider spacing between turbines in tilled landscapes relative to other locations.

The juxtaposition between wider turbine spacing yet less land transformation could influence decisions about the optimal location of new wind development. Our results suggest that for a fixed generating capacity, a facility in an agriculture landscape would require a greater amount of total area (if measured as a polygon around the turbines), yet produce less new land transformation relative to the same capacity in a different land-use type.

Our main argument is that geographic variation in land transformation exists and should be considered when forecasting or planning for wind energy, we hope that other researchers and policy makers will consider using more than a single point estimate of land transformation for wind energy. Our estimate for entire facilities (0.93 ha/MW) was higher than all but two previously reported values (1 ha/MW by the Bureau of Land Management (BLM) [23], and 2.4 ha/MW by Denholm [13], Table 1). The estimate from BLM was considered a high estimate, whereas the estimate from Denholm represented the maximum value in their data. Our highest values were 4.3 and 2.3 ha/MW. If one assumes direct measurements of land transformation are more accurate than estimates based on build out models, environmental impact reports, or derivations based on turbine spacing, then the majority of point estimates used in planning may have underestimated land transformation associated with new facilities.

Our work has implications for studies attempting to forecast future land transformation associated with wind development. Given the high levels of variation in land transformation reported here and elsewhere, using a single point estimate of land transformation for broad-scale generalization will likely produce vague and perhaps even misleading results, particularly when comparing land transformation across energy types. For example, using model-averaged coefficients for tilled landscapes to project new land transformation for 100GW of wind energy predicts 686 km$^2$ whereas using coefficients for forests yields 4606 km$^2$. We suggest using model coefficients that relate geographic variables to land transformation, in conjunction with geospatial map layers, to forecast land transformation from wind and other forms of energy production.

## 9.5 CONCLUSIONS

Ultimately, our results suggest the geographic context where energy is developed will play a fundamental role in the levels of actual or forecasted impact. This makes comparisons across energy types difficult, as we must control for underlying differences in pre-development factors that could influence impacts such as land cover, quantities and qualities of water and biodiversity, or even the perceived aesthetic value of a location. Such analyses will become complex if the impacts for different energy types changes with the geographic setting. For example, wind energy benefits on health and $CO_2$ reduction caused by displacing fossil fuels varies considerably across the U.S. [24]. Though adding complexity, variation across landscapes in variables that drive the impacts of energy development ultimately creates opportunities to plan national energy strategies that minimize impacts while maximizing benefits.

## REFERENCES

1. Hand MM, Baldwin S, DeMeo E, Reilly JM, Mai T, et al., editors (2012) Renewable electricity futures study (entire report). NREL/TP-6A20-52409 ed. Golden, CO: National Renewable Energy Laboratory.
2. Snyder B, Kaiser MJ (2009) Ecological and economic cost-benefit analysis of offshore wind energy. Renewable Energy 34: 1567–1578. doi: 10.1016/j.renene.2008.11.015
3. Pimentel D, Herz M, Glickstein M, Zimmerman M, Allen R, et al. (2002) Renewable energy: Current and potential issues. BioScience 52: 1111–1120. doi: 10.1641/0006-3568(2002)052[1111:recapi]2.0.co;2
4. McDonald RI, Fargione J, Kiesecker J, Miller WM, Powell J (2009) Energy sprawl or energy efficiency: Climate policy impacts on natural habitat for the united states of america. PLoS ONE 4: e6802. doi: 10.1371/journal.pone.0006802
5. Fthenakis V, Kim HC (2009) Land use and electrcity generation: A life-cycle analysis. Renewable and Sustainable Energy Reviews 13: 1465–1474. doi: 10.1016/j.rser.2008.09.017
6. Kiesecker JM, Copeland H, Pocewicz A, Nibbelink N, McKenney B, et al. (2009) A framework for implementing biodiversity offsets: Selecting sites and determining scale. BioScience 59: 77–84. doi: 10.1525/bio.2009.59.1.11
7. Fargione J, Kiesecker J, Slaats MJ, Olimb S (2012) Wind and wildlife in the northern great plains: Identifying low-impact areas for wind development. PLoS ONE 7: e41468. doi: 10.1371/journal.pone.0041468
8. Pacala S, Socolow R (2004) Stabilization wedges: Solving the climate problem for the next 50 years with current technologies. Science 305: 968–972. doi: 10.1126/science.1100103
9. Department of Energy (2008) 20% wind energy by 2030. Increasing wind energy's contribution to U. S. Electricity supply. DOE/GO-102008-2567. 248 p.
10. Energy Information Administration (2009) The national energy modeling system: An overview 2009. Washington D.C.: Department of Energy. DOE/EIA-0581 DOE/EIA-0581. 77 p.
11. Greenblatt J (2009) Clean energy 2030:Google's proposal for reducing U.S. dependence on fossil fuels [internet]. Version 170..
12. Kuvlesky WP, Brennan LA, Morrison ML, Boydston KK, Ballard BM, et al. (2007) Wind energy development and wildlife conservation: Challenges and opportunities. The Journal of Wildlife Management 71: 2487–2498. doi: 10.2193/2007-248
13. Denholm P, Hand M, Jackson M, Ong S (2009) Land-use requirements of modern wind power plants in the United States. NREL/TP-6A2-45834. 40 p.

14. Kiesecker JM, Evans JS, Fargione J, Doherty K, Foresman KR, et al. (2011) Win-win for wind and wildlife: A vision to facilitate sustainable development. PLoS ONE 6: e17566. doi: 10.1371/journal.pone.0017566

15. McCullagh P, Nelder JA (1989) Generalized linear models. Boca Raton, Fl: Chapman and Hall/CRC.

16. Elliott LP, Brook BW (2007) Revisiting Chamberlin: Multiple working hypotheses for the 21st century. BioScience 57: 608–614. doi: 10.1641/b570708

17. Chamberlin T (1890) The method of multiple working hypotheses. Science ns-15: 92–96. doi: 10.1126/science.ns-15.366.92

18. Anderson DR, Burnham KP, Thompson WL (2000) Null hypothesis testing: Problems, prevalence, and an alternative. The Journal of Wildlife Management 64: 912–923. doi: 10.2307/3803199

19. Burnham KP, Anderson DR (2002) Model selection and multimodel inference: A practical information-theoretic approach. New York: Springer-Verlag. 488 p.

20. Burnham KP, Anderson DR (1992) Data-based selection of an appropriate biological model:The key to modern data analysis. In: McCullough DR, Barrett RH, editors. Wildlife 2001:Populations. London, England: Elsevier. pp. 16–30.

21. Anderson DR, Link WA, Johnson DH, Burnham KP (2001) Suggestions for presenting the results of data analyses. The Journal of Wildlife Management 65: 373–378. doi: 10.2307/3803088

22. R Development Core Team (2013) R: A language and environment for statistical computing. R foundation for statistical computing, vienna, austria. Isbn 3-900051-07-0.

23. Bureau of Land Management (2005) Final programmatic environmental impact statement on wind energy development on blm-administered lands in the western United States. FES05-11. United States Department of Interior. Available at http://windeis.Anl.Gov/documents/fpeis/index.Cfm.

24. Siler-Evans K, Azevedo IL, Morgan MG, Apt J (2013) Regional variations in the health, environmental, and climate benefits of wind and solar generation. Proceedings of the National Academy of Sciences.

25. Elliot DL, Wendell LL, Gower GL (1991) An assessment of the available windy land area and wind energy potential in the contiguous United States. PNL-7789. 86 p.

26. Mackay DJ (2008) Sustainable energy - without the hot air. Cambridge: UIT. 366 p.

*There are several supplemental files that are not available in this version of the article. To view this additional information, please use the citation on the first page of this chapter.*

## CHAPTER 10

# Wind Turbines and Human Health

LOREN D. KNOPPER, CHRISTOPHER A. OLLSON,
LINDSAY C. MCCALLUM, MELISSA L. WHITFIELD ASLUND,
ROBERT G. BERGER, KATHLEEN SOUWEINE,
AND MARY MCDANIEL

## 10.1 INTRODUCTION

Wind power has been harnessed as a source of energy around the world for decades. Reliance on this form of energy is increasing. In 1996, the global cumulative installed wind power capacity was 6,100 MW; in 2011, that value had grown to 238,126 MW and at the end of 2013 it was 318,137 MW (1). While public attitude is generally overwhelmingly in favor of wind energy, this support does not always translate into local acceptance of projects by all involved (2). Opposition groups point to a number of issues concerning wind turbines, and possible effects on human health is one of the most commonly discussed. Indeed, a small proportion of people that live near wind turbines have reported adverse health effects such as (but not limited to) ringing in ears, headaches, lack of concentration, vertigo, and sleep disruption that they attribute to the wind turbines. This collection of effects has received the colloquial name "Wind Turbine Syndrome" (3).

*Wind Turbines and Human Health © Knopper LD, Ollson CA, McCallum LC, Whitfield Aslund ML, Berger RG, Souweine K, and McDaniel M.* Frontiers in Public Health *2,63 (2014). doi: 10.3389/ fpubh.2014.00063. Licensed under Creative Commons Attribution License, http://creativecommons. org/licenses/by/3.0/.*

**TABLE 1:** General summary of reviewed articles.

| General topic | Authors | Source | Keywords | General summary |
|---|---|---|---|---|
| Audible noise | Shepherd et al.(23) | Noise and Health | Health-related quality of life (HRQOL) | Cross-sectional study involving question-naires about quality of life living near and away from turbines. Statistically significant differences were noted in some HRQOL scores; residents within 2 km of a turbine reporting lower overall quality of life, physical quality of life, and environmental quality of life |
| | Janssen et al. (24) | Journal of the Acousti-cal Society of America | Annoyance, economic benefit, sen-sitivity, visual cues | Expanded on the data sets collected by Pedersen and Persson Waye (4, 5) and Pedersenetal. (17) in Sweden and the Netherlands. Authors evaluated self-reported annoyance indoors and outdoors compared to sound levels (Lden) from wind turbines. Like the authors before them who relied on these datasets, found that annoyance decreased with economic benefit and may have increased with noise sensitivity, visibility, and age. In comparison to other sources of environ-mental noise, annoyance due to wind turbine noise was found at relatively low noise exposure levels |
| | Verheijen-etal. (25) | Science of the Total Environment | Annoyance, noise limits | Objective was to assess proposed Dutch standards for wind turbine noise and consequences for people and feasibil-ity of meeting energy policy targets. Authors used a combination of audible and low-frequency noise models and functions to predict existing level of severely annoyed people living around existing wind turbines in the Nether-lands. Found that at 45dB (Lden) severe annoyance due to low-frequency noise unlikely; suggested that this noise limit is suitable as a trade-off between the need for protection against noise annoyance and the feasibility of national targets for renewable energy |

**TABLE 1:** *Cont.*

| General topic | Authors | Source | Keywords | General summary |
|---|---|---|---|---|
| | Bakkeretal. (26) | Science of the Total Environment | Annoyance, distress, economic benefit, sleep disturbance | A dose–response relationship was found between immission levels of wind turbine sound and self-reported noise annoyance. Sound exposure was also related to sleep disturbance and psychological distress among those who reported that they could hear the sound, however not directly but with noise annoyance. Respondents living in areas with other background sounds were less affected than respondents in quiet areas.Found that people, animals, traffic and mechanical sounds were more often identified as a source of sleep disturbance than wind turbines |
| | Nissenbaum et al.(27) | Noise and Health | Epworth Sleepiness Score (ESS), Pittsburgh Sleep Quality Index(PSQI), SF36v2 | Purpose of the investigations was to determine the relationship between reported adverse health effects and wind turbines among residents of two rural communities. Participants living 375–1,400 m and 3.3–6.6 km were given questionnaires to obtain data about sleep quality, daytime sleepiness and general physical and mental health. Authors reported that when compared to people living further away than 1.4 km from wind turbines, those people living within 1.4 km of wind turbines had worse sleep, were sleepier during the day and had worse mental health scores |
| | Ollson etal. (28) | Noise and Health | Rebuttal to Nissenbaum et al.(27) | Suggested that Nissenbaum et al. (27) extended their conclusions and discussion beyond the statistical findings of their study and that they did not demonstrate a statistical link between wind turbines–distance–sleep quality–sleepiness and health. In fact, their own statistical findings suggest that although scores may be statistically different between near and far groups for sleep quality and sleepiness, they are no different than those reported in the general population. The claims of causation by the authors (i.e., wind turbine noise) for negative scores are not supported by their data |

**TABLE 1:** *Cont.*

| General topic | Authors | Source | Keywords | General summary |
|---|---|---|---|---|
| | Barnard(29) | Noise and Health | Rebuttal to Nissenbaum et al.(27) | Pointed out a number of problems with Nissenbaumetal.(27) study and suggested that data presented do not justify the very strong conclusions reached by the authors |
| | Mroczeketal. (30) | Annals of Agricultural and Environmental Medicine | SF-36,Visual Analog Scale (VAS) | Purpose of study was to assess how people's quality of life is affected by the close proximity of wind farms. Authors found that close proximity of wind farms does not result in the worsening of the quality of life based on the Norwegian version of the SF-36 General Health Questionnaire, the Visual Analog Scale (VAS) for health assessment, and original questions |
| | Taylor et al. (31) | Personality and Individual Differences | Personality-traits | Study examine the influence of negative oriented personality (NOP) traits on the effects of wind turbine noise and reporting non-specific symptoms (NSS). Results of the study showed that while calculated actual wind turbine noise did not predict reported symptoms, perceived noise did. |
| | Evans and Cooper (32) | Acoustics Australia | Predicted and measured noise levels | A comparison of predicted noise levels from founr commonly applied prediction methods against measured noise levels from six operational wind farms (at 13 locations) in accordance with the applicable guidelines in South Australia. Results indicate that the methods typically over-predicted wind farm noise levels but that the degree of conservatism appeared to depend on the topography between the wind turbines and the measurement location. |
| | Maffeietal. (33) | International Journal of Environmental Research and Public Health | Visual cues, perception | Investigated the effects of the visual impact of wind turbines on the perception of noise. Found distance was a strong predictor of an individual's reaction to the wind farm; data showed that increased distance resulted in a more positive general evaluation of the scenario and decreased perceived loudness, noise annoyance, and stress caused by sound. Found the color of the wind turbines (base and blade stripes) impacted an individual's perception of noise. |

**TABLE 1:** *Cont.*

| General topic | Authors | Source | Keywords | General summary |
|---|---|---|---|---|
| | Van Rent-erghem et al.(34) | Science of the Total Environment | Annoyance, attitude, laboratory experiment, visual cues | Conducted a two-stage listening experiment to assess annoyance, recognition, and detection of noise from a single wind turbine. Results support the hypothesis that non-noise variables, such as attitude and visual cues, likely contributed to the observation that people living near wind turbines (who did not receive an economic benefit from the turbines) report higher levels of annoyance at lower sound pressure levels than would be predicted for other community noise sources. |
| | Baxter et al. (35) | Energy Policy | Risk perception, economic benefit, community conflict, policy | Conducted a study to investigae the role of health risk perception, economic benefit, and community conflict on wind turbine policy. Two communities were assessed: one located in proximity to two operating wind farms and a control community without turbines. Authors found that residents from the community with operational wind energy projects were more supportive of wind turbines than residents in the area without turbines. |
| | Chapman et al. (6) | PLoS One | Psychogenic effects, nocebo, community complaints | Provided an overview of the growing body of literature supporting the notion that the attribution of symptoms and disease to wind turbine exposure is a modern health worry. Suggested that nocebo effects likely play an important role in the observed increase in wind farm-related health xomplaints. Suggested that reported historical and geographical variations in complaints were consistent with "communicated diseases" with nocebo effects likely to play an important role in the etiology of complaints rather than direct effects from turbines. |

**TABLE 1:** *Cont.*

| General topic | Authors | Source | Keywords | General summary |
|---|---|---|---|---|
| | Whitfield Aslund et al. (36) | Energy Policy | Predicted annoyance, modeling | Used previously reported dose-response relationships between wind turbine noise and annoyance to predict the level of community noise annoyance that may occur in the province of Ontario. The results of this analysis indicated that the current wind turbine noise restrictions in Ontario will limit community exposure to wind turbine-related noise such that levels of annoyance are unlikely to exceed previously established background levels of noise-related annoyance from other common noise sources. |
| Low-frequency noise and infrasound | Møller and Pedersen (37) | Journal of the Acoustical Society of America | Annoyance, insulation, indoor sound levels | Conducted a low-frequency noise study from four large turbines (>2 MW) and 44 other small and large turbines (7>2 MW and 37>2 MW). Low-frequency sound insulation was measured for 10 rooms under normal living conditions in houses exposed to low-frequency noise. Concluded that the spectrum of wind turbine noise moves down in frequency with increasing turbine size. Suggested that the low-frequency part of the noise spectrum plays an important role in the noise at neighboring properties. They hypothesized that if the noise from the investigated large turbines had an outdoor level of 44 dB (A) there was a risk that a substantial proportion of the residents would be annoyed by low-frequency noise, even indoors. |
| | Bolinetal. (38) | Environmental Research Letters | Health effects, review, turbulence | Conducted a literature review over a 6-month period ending April 2011 into the potential health effects related to infrasound and low-frequency noise exposure surrounding wind turbines. Concluded that empirical support was lacking for claims that low-frequency noise and infrasound causes serious health affects in the form of "vibracoustic disease," "wind turbine syndrome," or harmful efects on the inner ear. |

**TABLE 1:** *Cont.*

| General topic | Authors | Source | Keywords | General summary |
|---|---|---|---|---|
| | Randetal. (39) | Bulletin of Science, Technology and Society | Indoor sound levels,health effects,acute effects | Studies took places over a 2-day period inside a home where people were self-reporting serious adverse health effects. Authors reported on wind speed at hub of turbine, dB (A) and dB (G) filtering indoors and outdoors. Reported on acuted effects. |
| | Ambrose et al. (40) | Bulletin of Science, Technology and Society | | |
| | Turnbull et al. (41) | Acoustics Australia | Underground measurement, comparative study | Developed an underground technique to measure infrasound. Measured infrasound at two Australian wind farms as well as in the vicinities of a beach, a coastal cliff, the city of Adelaide, and a power station. Reported that the measured levels at wind farms below the audibility threshold and similar to that of urban and coastal environments and near other engineered noise sources. Level of infrasound from wind farms at 360 and 85 m [61 and 72 dB (G), respectively] was comparable to that observed data distance of 25 m from ocean waves [75 dB (G)] |
| | Crichton et al. (7) | Health Psychology | Negative expectations, symptom reporting, laboratory experiment | Examined the possibility that expecta-tions of negative health effects from exposure to infrasound promotes symptom reporting using a sham con-trolled, double-blind provocation study. Participants in the high-expectancy group reported significant increases in the num-ber and intensity of symptoms experi-enced during exposure to both infrasound and sham infrasound. Conversely, there were no symptomatic changes in the low-expectancy group. |

**TABLE 1:** *Cont.*

| General topic | Authors | Source | Keywords | General summary |
|---|---|---|---|---|
| | Crichton et al. (8) | Health Psychology | Negative and positive expectations, symptom reporting, laboratory experiment | Authors investigated how positive expectations can produce a reduction in symptoms. Expectations were found to significantly alter symptom reporting: participants who were primed with negative expectations became more symptomatic overtime, suggesting that their experiences during the first exposure session reinforced expectations and led to heightened symptomatic experiences in subsequent sessions. |
| Electromagnetic fields | Havas and Colling (42) | Bulletin of Science, Technology and Society | Poor power quality, ground current, electrical hypersensitivity | Authors hypothesized that symptoms of some living near wind turbines could be caused by electromagnetic waves in the form of poor power quality (dirty electricity) and ground current resulting in health effects in those that are electrically hypersensitive. Indicated that individuals reacted differently to both sound and electromagnetic waves and this could explain why not everyone experiences the same health effects living near turbines. |
| | Israel et al. (43) | Environmentalist | Vibration measurement, noise, risk | Conducted EMF, sound, and vibration measurements at wind energy parks in Bulgaria. Concluded that EMF levels were not of concern from wind farm |
| | McCallum et al.(44) | Environmental Health | Variable distances and wind, residential measures | Magnetic field measurements were collected in the proximity of 15 wind turbines, two substations, buried and overhead collector and transmission lines and nearby homes. Results sugges their is nothing unique to wind farms with respect to EMF exposure; in fact, magnetic field levels in the vicinity of wind turbines were lower than those produced by many common household electrical devices and were well below and existing regulatory guidelines with respect to human health. |
| Review articles, editorials and social commentaries | Bulletin of Science, Technology and Society (BSTS)Special Edition | Bulletin of Science, Technology and Society | Various authors, health effects, social commentary, opinion pieces | Special edition made up of nine articles devoted entirely to wind farms and potential health effets. Many of the articles in the special edition were written as opinion pieces or social commentaries. |

**TABLE 1:** *Cont.*

| General topic | Authors | Source | Keywords | General summary |
|---|---|---|---|---|
| | Hanning and Evans(45) | British Medical Journal | Sleep disturbance | Purpose was to opine on the relationship between wind turbine noise and health effects. Suggested that a large body of evidence exists to suggest that wind turbines disturb sleep and impair health that distances and external noise levels that are permitted in most jurisdictions. |
| | Chapman (46) | British Medical Journal | Weight of evidence | In a rebuttal to Hanning and Evans (45), Chapman points to 17 independent reviews of the literature around wind turbines and human health that contrast the opinion of Hanning and Evans. |
| | Farboud et al. (47) | Journal of Laryngology and Otology | Low-frequency noise (LFN), infrasound (IS), inner ear physiology, wind turbine syndrome | Conducted a literature search for articles published within the last 10 years, using the PubMed database and the Google Scholar search engine, to look at the effects of low-frequency noise and infrasound. Suggested that evidence available was incomplete and until the physiological effects of LFN and infrasound were fully understood, it was not possible to conclusively state that wind turbines were not causing any of the reported effects. |
| | McCubbin and Sovacool(48) | Energy Policy | Comparative study, natural gas, health, and environmental benefits | Compared the health and environmental benefits of wind power in contrast to natural gas. |
| | Roberts and Roberts (49) | Journal of Environmental Sciences | PubMed-based review, low-frequency noise (LFN), infrasound (IS), health effects | Conducted a summary of the peer-reviewed literature on the research that examined the relationship between human health effects and exposure to low-frequency sound and sound generated from the operation of wind turbines. Concluded that a specific health condition or collection of symptoms has not been documented in the peer-reviewed, published literature that has been classified as a "disease" caused by exposure to sound levels and frequencies generated by the operations of wind turbines. |

**TABLE 1:** *Cont.*

| General topic | Authors | Source | Keywords | General summary |
|---|---|---|---|---|
| | Chapman and St. George (50) | Australian and New Zealand Journal of Public Health | Vibroacoustic disease (VAD); factoid | Investifated the extent to which VAS and its alleged association with wind turbine exposure had received scientific attention, the quality of that association and how the alleged association gained support by wind farm opponents. Based on a structured scientific database and Google search strategy, the authors showed that VAD has received virtually no scientific recognition that there is no evidence of even rudimentary quality that vibroacoustic disease is associated with or caused by wind turbines. Stated that an implications of this "factoid"—defined as questionable or spurious statements—may have been contributing to nocebo effects among those living near turbines |
| | Jeffery et al. (51) | Canadian Family Physician | Health effects | Overall goal of these commentary pieces was to provide information to physicians regarding the possible health effects of exposure to noise produced by wind turbines and how these many manifest in patients. |
| | Jeffery et al. (52) | Canadian Journal of Rural Medicine | | |

The reason for the self-reported health effects is highly debated and information fueling this debate is found primarily in four sources: peer-reviewed studies published in scientific journals, government agency reports, legal proceedings, and the popular literature and internet. Some argue that reported health effects are related wind turbine operational effects [e.g., electromagnetic fields (EMF), shadow flicker from rotor blades, audible noise, low-frequency noise (LFN) and infrasound]; others suggest that when turbines are sited correctly, reported effects are more likely attributable to a number of subjective variables, including nocebo responses, where the etiology of the self-reported effect is in beliefs and expectations rather than a physiologically harmful entity (4–8). In 2011, Knopper and Ollson (9) published a review that contrasted the human health effects

that had been purported to be caused by wind turbines in popular literature sources with what had been reported in the peer-reviewed scientific
literature as well as by various government agencies. At that time, only 15
articles in the peer-reviewed scientific literature that specifically addressed
issues related to human health and wind turbines were available [i.e., (4,
5, 10–22)].

Based on their review, Knopper and Ollson (9) concluded that although
there was evidence to suggest that wind turbines can be a source of annoyance to some people, there was no evidence demonstrating a direct
causal link between living in proximity to wind turbines and more serious
physiological health effects. Furthermore, although annoyance has been
statistically significantly associated with wind turbine noise [especially at
sound pressure levels >40 dB(A)], a convincing body of evidence exists
to show that annoyance is more strongly related to visual cues and attitude
than to wind turbine noise itself. In particular, this was highlighted by the
fact that people who benefit economically from wind turbines (e.g., those
who have leased their property to wind farm developers) reported significantly lower levels of annoyance than those who received no economic
benefit, despite increased proximity to the turbines and exposure to similar
(or louder) sound levels.

In the years following the publication of Knopper and Ollson (9), the
debate surrounding the relationship between wind turbines and human
health has continued, both in the public and within the scientific community. In this review, we provide a bibliographic-like summary and
analysis of the science around this issue specifically in terms of noise
(including audible, LFN, and infrasound), EMF, and shadow flicker.
Stemming from this review, we provide weight of evidence conclusions
and a number of best practices for wind turbine development in the context of human health.

## 10.2 METHODS

The authors worked with a professional Health Sciences Information
Specialist to develop a search strategy of the literature. Combinations of
key words (i.e., annoyance, noise, environmental change, sleep distur-

bance, epilepsy, stress, health effect(s), wind farm(s), infrasound, wind turbines(s), LFN, EMF, wind turbine syndrome, neighborhood change) were entered into PubMed, the Thomson Reuters Web of KnowledgeSM and Google. No date restrictions were entered and literature was assessed up to the submission date of this manuscript (April 2014). The review was conducted in the spirit of the evaluation process outlined in the Cochrane Handbook for Systematic Reviews of Interventions.

As of the publication date of this review, there are close to 60 scientific peer-reviewed articles on the topic. Sources of information other than peer-reviewed scientific literature (e.g., websites, opinion pieces, conference proceedings, unpublished documents) were purposely excluded in this review because they are often unreliable and provide information that is typically anecdotal in nature or not traceable to scientific sources. A general summary, and key words of the articles reviewed herein, are presented in Table 1. These summaries provide results as they were reported by the authors of the articles and are without secondary interpretation.

Through the systematic review process, it was evident that there was significant variability in both the measures of exposure (i.e., proximity to turbines, field noise measures, lab noise measures, or magnetic field measurements) and the health outcomes examined (i.e., annoyance, sleep scores, and various quality of life metrics). The methodological heterogeneity in study designs across the selected health-based investigations inhibited a quantitative combination of results. In other words, meta-analytic methods were not appropriate for this updated systematic review of the literature on wind turbine and health effect. Rather qualitative interpretation is provided.

## 10.3 RESULTS

### 10.3.1 OVERALL NOISE

Knopper and Ollson (9) reviewed a number of studies that examined the noise levels produced by wind turbines, perception of wind turbine noise, and/or responses to wind turbine noise [e.g., (4, 5, 10, 12, 13, 15–18, 21)]. The results of more recent studies that investigated wind turbine noise with respect to potential human health effects are summarized below in chronological order of publication.

Shepherd et al. (23): Shepherd et al. reported on a cross-sectional study comparing health-related quality of life (HRQOL) of people living in proximity (i.e., <2 km) to a wind farm to a control group living >8 km away from the nearest wind farm. It involved self-administered questionnaires that included the World Health Organization (WHO) quality of life scale, in semi-rural New Zealand. The turbine group was drawn from residents of 56 homes in South Makara Valley, all within 2 km of a wind turbine. General outdoor noise levels in the area, obtained from a conference proceeding by Botha (53), were reported to range from 24 to 54 dB(A). The comparison group was taken from 250 homes in a geographically and socioeconomically matched area, at least 8 km from any wind farm in the region. General outdoor noise levels for the comparison group were not reported. The questionnaire was named the "2010 Well-being and Neighborhood Survey" in order to mask the true intent of the study and reduce bias against wind turbines. This is similar to the work of Pedersen in Europe, in that the surveys were not explicitly about wind turbines. Response rates were 34% from the Turbine group (number of participants n = 39) and 32% from the Comparison group (n = 158).

Overall, Shepherd et al. reported statistically worse (p < 0.05) scores in the Turbine group for physical HRQOL, environmental QOL and HRQOL in general. There was no statistical difference in social or psychological scores. Based on these results, the authors concluded that "utility-scale" wind energy generation was not without adverse health impacts on nearby residents and suggested setback distances need to be >2 km in hilly terrain. However, there are a number of limitations in this study that undermine the conclusion stated above. One key concern is that the results were based on only a limited number of participants (n = 39) for the Turbine group. In comparison, the survey datasets compiled in Sweden and the Netherlands by Pedersen and Persson Waye (4, 5) and Pedersen et al. (17), respectively, involved a total of 1,755 respondents overall. In these surveys, the only response found to be significantly related to A-weighted wind turbine noise exposure was annoyance, even though a number of physiological and psychological variables were also investigated. In addition, Shepherd et al. did not discuss the impact of participants' attitudes or visual cues that may have influenced the reports of decreased HRQOL. Given that other studies have indicated that annoyance was more closely related to visual cues and attitude, this could provide further explanation of why overall HRQOL scores were lower in the Turbine group. Presumably all residents within 2 km of a turbine would be able to see one, or more, of the turbines. Furthermore, although it was implied in the title of the article that noise from wind turbines was causing the observed effects, the study did not include either measured or estimated wind turbine noise exposure values for the individual survey respondents. Therefore, they were unable to demonstrate a dose–response relationship between the observed responses and exposure to wind turbine noise. In light of this, as recognized by Shepherd et al. (23), it is possible that the observed effects were driven by other causes such as conflicts between the community and the wind farm developers rather than a direct result of noise exposure. Based on the limitations discussed above, we consider that the authors' recommendation for a 2 km setback distance was not supported by the evidence presented in this study.

Janssen et al. (24): expanding on the datasets collected by Pedersen and Persson Waye (4, 5) and Pedersen et al. (17) in Sweden and the Netherlands, Janssen et al. evaluated self-reported annoyance indoors and outdoors compared to sound levels (Lden) from wind turbines. To derive the Lden, the authors added a correction factor of 4.7 dB(A) to outdoor A-weighted sound pressure levels from the datasets used in the previous studies. Annoyance in this study was ranked on a 4-point scale: 1 was "not annoyed," 2 was "slightly annoyed," 3 was "rather annoyed," and 4 was "very annoyed." Visual cue ("Can you see a wind turbine from your dwelling or your garden/balcony?"), economic benefit ["Are you a (co) owner of one or more wind turbines?"], and noise sensitivity (on either a 4 or 5 point scale with 1 representing "not sensitive" and 4 or 5 representing "very/extremely sensitive") were also assessed. Like the authors before them who relied on these datasets, Janssen et al. found that annoyance decreased with economic benefit and may have increased with noise sensitivity, visibility, and age. Rates of annoyance indoors from wind turbines to industrial noise from stationary sources and air, road and rail noise were also compared and it was concluded that: "...annoyance due to wind turbine noise is found at relatively low noise exposure levels" and that "some similarity is found in the range Lden 40–45 dB between the percentage of annoyed persons by wind turbine noise and aircraft noise."

Verheijen et al. (25): the objective of this study was to assess the proposed Dutch protective standards for wind turbine noise, both on consequences for inhabitants and feasibility of meeting energy policy targets. The authors used a combination of audible and LFN models and functions derived by Janssen et al. (24) to predict the existing level of severely annoyed people living around existing wind turbines in the Netherlands. They estimated that there were approximately 1,500 severely annoyed individuals, in a total population of approximately 440,000 living at sound levels of 29 dB(Lden) around wind turbines. The authors reported that: "For The Netherlands, a socially acceptable percentage of severely annoyed lies around 10%, which can be derived from the existing limits and dose–response functions of railway and road noise. This would result in an acceptable noise reception limit for wind turbines of about 47 to 49 dB." The authors decided to examine the feasibility of lowering the limit below 47–49 dB(Lden). They estimated that it may be feasible from a land mass perspective to lower the noise limit to 40 dB(Lden); however, given that lands are often rejected due to reasons other than noise that another value should be selected. They stated "The percentage of severely annoyed at 45 dB is rated at 5.2% for wind turbine noise, which is well below 10% that corresponds to the existing road and railway traffic noise limits." They also determined that, at 45 dB(Lden), severe annoyance effects due to LFN were unlikely and suggested that this noise limit suited as a trade-off between the need for protection against noise annoyance and the feasibility of national targets for renewable energy.

Bakker et al. (26): the purpose of this study was to evaluate the relationship between exposure to the sound of wind turbines and annoyance, self-reported sleep disturbance, and psychological distress of people that live in their vicinity. This investigation relied on survey data, previously reported and discussed by Pedersen et al. (17), collected from 725 residents of the Netherlands living in the vicinity of wind turbines. As reported by Pedersen et al. (17), survey respondents answered questions about environmental factors and road traffic noise (and wind noise) as well as the effect of wind turbines on annoyance, sleep disturbance, and psychological distress.

Bakker et al. differed from Pedersen et al. (17) in that it provided a direct comparison of people who economically benefited from turbines with those who did not, specifically in relation to annoyance. Bakker et al. (26) reported that only 3% of survey respondents receiving economic benefit from wind turbines reported being "rather annoyed" or "very annoyed" by wind turbine noise when outdoors, while none reported being rather or very annoyed by wind turbine noise when indoors. In comparison, the proportions of survey respondents who did not receive an economic benefit who reported being rather or very annoyed indoors and outdoors were 12 and 8%, respectively, even though they were exposed to significantly lower levels of wind turbine sound.

What is more, Bakker et al. also compared sound-related sources of sleep disturbance in rural and urban areas in respondents who did not benefit economically from wind turbines. They found that people, animals, traffic, and mechanical sounds were more often identified as a source of sleep disturbance than wind turbines. In fact, in rural areas, only 6% of people identified wind turbines as the sound source of sleep disturbance compared to 11.7% for people/animals and 12.5% for traffic/mechanical sounds. In urban areas, only 3.8% of people identified wind turbines as the sound source of sleep disturbance compared to 14.4% for people/animals and 16.9% for traffic/mechanical sounds.

Nissenbaum et al. (27), Ollson et al. (28), and Barnard (29): the stated purpose of the investigations conducted by Nissenbaum et al. was to determine the relationship between reported adverse health effects and wind turbines among residents of two rural communities. Participants living 375–1,400 m and 3.3–6.6 km were given questionnaires to obtain data about sleep quality [using the Pittsburgh Sleep Quality Index (PSQI)], daytime sleepiness [using the Epworth Sleepiness Score (ESS)], and general physical and mental health (MH) (using the SF36v2 health survey). Overall, the authors reported that when compared to people living further away than 1.4 km from wind turbines, those people living within 1.4 km of wind turbines had worse sleep, were sleepier during the day, and had worse MH scores. Based on these findings the authors concluded that: "...the noise emissions of IWTs disturbed the sleep and caused daytime sleepiness and impaired mental health in residents living within 1.4 km of the two IWT installations studied."

In a subsequent issue of Noise and Health, two letters to the editor were published that were critical of this study and its conclusions (28, 29). In particular, the letter from Barnard (29) criticized the statistical analysis in Nissenbaum et al. (27), which stated that there was a "strong" dose–response relationship between distance to the nearest wind turbine and both the "PSQI" and the "Epworth Sleepiness Scale." Barnard stated: "I cannot see how this is justified, given the presented data. In contrast to the conclusions, Figure 1 and Figure 2 in the paper... show a very weak dose-response, if there is one at all. The near horizontal 'curve fits' and large amount of 'data scatter' are indications of the weak relationship between sleep quality and turbine distance. The authors seem to use a low P value as a support for the hypothesis that sleep disturbance is related to turbine distance. A better interpretation of the P value related to a near horizontal line fit would be that it suggests a high probability of a weak-dose response. Correlation coefficients are not given, but should have been given, to indicate the quality of the curve fits." Ollson et al. (28) pointed out that Nissenbaum et al. extended their conclusions and discussion beyond the statistical findings of their study. They stated "We believe that they have not demonstrated a statistical link between wind turbines–distance–sleep quality–sleepiness and health. In fact, their own statistical findings suggest that although, scores may be statistically different between near and far groups for sleep quality and sleepiness, they are not different than those reported in the general population. The claims of causation by the authors (i.e., wind turbine noise) for negative MCS scores are not supported by their data. This work is exploratory in nature and should not be used to set definitive setback guidelines for wind-turbine installations."

Mroczek et al. (30): Mroczek et al. published the results of a study conducted in 2010 that evaluated the impact of living in close proximity to wind turbines on an individual's perceived quality of life. The study group consisted of 1,277 randomly selected Polish adults (703 women and 574 men) living in the vicinity of wind farms. The different distance (house to turbine) groups were: <700 m, from 700 to 1000 m, from 1,000 to 1,500 m, and >1,500 m. The quality of life was measured using the Norwegian version of the SF-36 General Health (GH) Questionnaire, the Visual Analog Scale (VAS) for health assessment, and some original questions about approximate distance to wind farm, age, gender, education, and profession. The SF-36 (Short Form 36) Questionnaire consists of 36 questions divided into 8 subscales: physical functioning (PF), role functioning physical (RP), bodily pain (BP), GH, vitality (V), social functioning (SF), role functioning emotional (RE), MH, and one additional question regarding health changes.

According to the authors "The respondents assessed their health through answering questions included in the SF-36 and VAS. They were asked to mark the point corresponding with their well-being on the level from 0 to 100, where 0 denoted the worst possible state of health and 100 – excellent health." The results showed that regardless of the distance from the wind farm (i.e., from <700 to >1,500 m) respondents ranked their PF scores as highest out of all of the quality of life components. Overall, people living closest to wind farms assessed their quality of life as higher than those living in more distant areas. The scores for the MH component, GH, SF, and RE were highest in the group living closest to the wind farms and lowest by those living greater than 1.5 km away. The authors noted that there may have been confounding factors that contributed to the observed results (e.g., economic factors). Since other studies have shown links between self-reported health status, proximity to wind turbines and the direct influence of economic benefit on levels of annoyance [e.g., (17, 26)], these major confounding factors also need to be considered when interpreting the results of the Mroczek et al. study on quality of life and proximity to wind turbines.

Taylor et al. (31): this study examined the influence of negative oriented personality (NOP) traits on the effects of wind turbine noise and reporting on non-specific symptoms (NSS). The study was conducted based on the hypothesis that the public has become increasingly concerned with attributing NSS to environmental features (e.g., wind turbines). The study focused on three NOP traits in particular: neuroticism (N), negative affect (NA), and frustration intolerance (FI). The authors noted that previous research has demonstrated that individuals with high N and NA typically evaluate their environment more negatively. Furthermore, FI may have impacted the way an individual perceived and evaluated environmental factors from an inability to bear or cope with perceived negative emotions, thoughts and events. A survey was mailed out to 1,270 households within 500 m of eight 0.6 kW turbine installations and within 1 km of four

5 kW turbines in two cities in the U.K. Individuals within the household (>18 years old) could anonymously complete the survey and mail the results back or submit them online. In total, 138 completed surveys were returned. Actual sound levels were calculated for those households who completed the survey, and participants were asked to describe the perceived noise, including the type of noise (e.g., swooshing, whistling, buzzing), frequency, and loudness (based on a 0–4 ranking scale). Participants were also asked a series of questions to determine the level of NOP traits and related health/symptom reporting information.

The results of the study showed that while calculated actual wind turbine noise did not predict reported symptoms, perceived noise did. Specifically: "…for those higher in NOP traits, there was a stronger link between perceived noise and symptom reporting. There was however, no relationship between calculated actual noise from the turbine and participants attitude to wind turbines. This means that those who had a more negative attitude to wind turbines perceived more noise from the turbine, but this effect was not simply due to individuals being able to actually hear the noise more."

Evans and Cooper (32): in their paper called "Comparison of predicted and measured wind farm noise levels and implications for assessments of new wind farms," Evans and Cooper present a comparison of predicted noise levels from four commonly applied prediction methods against measured noise levels from six operational wind farms (conducted at 13 locations) in accordance with the applicable guidelines in South Australia. The results indicate that the methods typically over-predicted wind farm noise levels but that the degree of conservatism appeared to depend on the topography between the wind turbines and the measurement location. Briefly, Evans and Cooper found that the commonly used ISO 9613-2 model (with completely reflective ground) and the CONCAWE model generally over-predicted noise levels by 3–6 dB(A), but the amount of over-prediction was related to the topography (i.e., relatively flat topography or a steady slope from the turbines). However, at sites where there was a significant concave slope from the turbines down to the measurement sites, these commonly used prediction methods were typically accurate, with the potential of marginal under-prediction in some cases (when ISO 9613-2 used 50% absorptive ground).

A requirement of many regulatory agencies is that noise modeling be conducted by developers prior to the construction of wind turbines. A common criticism of this approach is that modeled values are not representative of actual noise from operational wind farms. Evans and Cooper's findings show that this is not the case, but caution about the role of topography.

Maffei et al. (33): despite the fact that wind farms are represented as environmentally friendly projects, wind turbines are viewed by some as visual and audible intruders that spoil the landscape and generate noise. Consequently, Maffei et al. (33) conducted a study investigating the effects of the visual impact of wind turbines on the perception of noise. The study consisted of 64 participants (34 males, 30 females) who resided in either urban or rural areas. Participants were asked to fill out a questionnaire to obtain information regarding age, gender, education, and local neighborhood characteristics. A number of statements were then submitted to the participants where they were asked to respond based on a 100-point Likert scale ranging from "disagree strongly" to "agree strongly." The statements were based on personal views about green energy, wind turbines, noise, and other related subject matter. Subsequently, a virtual reality scenario was created to emulate the visual impact of a wind farm on a rural landscape and included an audio component recorded from a 16 turbine wind farm in Frigento, Italy. In total, three factors were manipulated in the experiment: distance from the wind farm (150, 250, and 500 m); the number of wind turbines (1, 3, and 6); the color of the base of the turbine and any stripes on the blades (white, red, brown, green). Each participant was asked to view all of the scenarios using a 3D visor and asked to respond to a number of questions pertaining to perceived loudness, sound pleasantness, noise annoyance, sound stress, sound tranquility, and visual pleasantness.

The results found that distance was a strong predictor of an individual's reaction to the wind farm. In particular, the data showed that increased distance resulted in a more positive general evaluation of the scenario and decreased perceived loudness, noise annoyance, and stress caused by sound. Additionally, the authors found that the color of the wind turbines (base and blade stripes) impacted an individuals' perception of noise. Generally, white and green turbines were preferred to brown and red ones. Specifically, green turbines scored the highest since they were perceived as being the "most integrated" into the landscape. The authors concluded that their results confirmed the interconnectedness between auditory and visual components of individual perception.

Van Renterghem et al. (34): Van Renterghem et al. (34) conducted a two-stage listening experiment to assess annoyance, recognition, and detection of noise from a single wind turbine. A total of 50 participants with "normal" hearing abilities participated in the experiment and were classified as having a positive to neutral attitude toward renewable energy. In situ recordings made at close distance (30 m downwind) from a 1.8 MW turbine operating at 22 rotations per minute (rpm) were mixed with road traffic noise and processed to simulate indoor sound pressure levels at 40 dB(LAeq). In the first stage, where participants were unaware of the true purpose of the experiment, samples were played during a quiet leisure activity. Under these conditions (i.e., when people were unaware of the different sources of noise), pure wind turbine noise produced similar annoyance ratings as unmixed highway noise at the same equivalent level, while annoyance from local road traffic was significantly higher. These results supported the hypothesis that non-noise variables, such as attitude and visual cues, likely contributed significantly to the observation that people living near wind turbines (who do not receive an economic benefit from the turbines) report higher levels of annoyance at lower sound pressure levels than would be predicted for other community noise sources [e.g., (17, 24)].

In the second stage of the Van Renterghem et al. (34) study, participants were allowed to listen to a recording of unmixed wind turbine sound [at 40 dB(A)] for 30 s in order to familiarize themselves with the sound. After this, they listened to 10 sets of paired sound samples; one of which contained unmixed road traffic noise and the other that contained wind turbine noise mixed with road traffic at signal-to-noise ratios varying between −30 dB(A) and +10 dB(A). For each pair, participants were asked to identify which of the two samples contained the wind turbine noise. The detection of wind turbine noise in the presence of highway noise was found a "signal-to-noise" ratio as low as −23 dB(A). This demonstrated that once the subject was familiar with wind turbine noise, it could easily be detected even in the presence of highway traffic noise. This could also help explain the increased rates of noise annoyance at home reported by

Pedersen et al. (17) and Janssen et al. (24) since residents would be familiar with the sound and be able to discern it if they listened for it when primed by visual cues. Overall, the findings support the idea that noticing the sound could be an important aspect of wind turbine noise annoyance. Awareness of the source and recognition of the wind turbine sound was also linked to higher levels of annoyance. Van Renterghem et al. noted that: "The experiment reported in this paper supports the hypothesis that previous observations, reporting that retrospective annoyance for wind turbine noise is higher than that for highway noise at the same equivalent noise level, is grounded in higher level appraisal, emotional, and/or cognitive processes. In particular, it was observed that wind turbine noise is not so different from traffic noise when it is not known beforehand."

Baxter et al. (35): in 2010, Baxter and colleagues conducted a study to investigate the role of health risk perception, economic benefit, and community conflict on wind turbine policy. The study, published in 2013, had two parts: a literature review and quantitative survey meant to determine perceptions of wind turbines and how they are linked to support or opposition to wind turbines in the community. Two communities were assessed: one located in proximity to two operating wind farms and a control community without turbines. Overall, the authors found that residents from the community with operational wind energy projects (which were introduced prior to the Green Energy Act in Ontario) were more supportive of wind turbines than residents in the area without turbines (78 vs. 29%, with "support" defined as agreeing to vote in favor of local turbines). The authors also reported that residents in the turbine community were more accepting of turbine esthetics than people in the control community and less worried about health impacts, this despite the fact that the wind farms in the "case" group were in some cases closer to homes than currently permitted.

Baxter et al. indicated that the lack of support in the control community could have been due to political lobbying during the provincial election, where one candidate suggested a moratorium on wind turbine as part of their campaign. The authors also highlighted the role of health risk perception (which seemed linked to political lobbying) as a variable leading to the lack of support. The finding that "Our study highlights the need to add health risk perception to the agenda for social research on turbines" is valid, albeit dated in the Ontario context, since an integral part of any wind development project in Ontario is public consultation with wind turbines and health as a fundamental component. These findings supported the idea that perception of health risks is heavily impacted by expectation, media coverage, and that "hands on experience" could serve to increase familiarity and decrease concerns.

Chapman et al. (6): the authors provided an overview of the growing body of literature supporting the notion that the attribution of symptoms and disease to wind turbine exposure is a modern health worry. Chapman et al. also suggested that nocebo effects likely play an important role in the observed increase in wind farm-related health complaints. By evaluating records of complaints from wind farm companies about noise or health from residents living near 51 wind farms across Australia, two theories about the etiology of complaints were tested: one being direct effects from turbines and the other being "psychogenic" effects brought on by nocebo effects.

Chapman et al. found a number of historical and geographical variations in wind farm complaints from Australians.

1. Nearly 65% of Australian wind farms, 53% of which have turbines >1 MW, have never been subject to noise or health complaints. These farms have an estimated 21,633 residents within 5 km and have operated complaint-free for a cumulative 267 years. No complaints were reported in Western Australia and Tasmania.
2. One in 254 residents across Australia appeared to have ever complained about health and noise, and 73% of these residents live near 6 wind farms that have been targeted by anti-wind farm groups. Ninety percentage of complaints were made after anti-wind farm groups added health concerns to their wider opposition in 2009.
3. In the years after, health or noise complaints were rare despite large and small-turbine wind farms having operated for many years.

It was suggested that reported historical and geographical variations in complaints were consistent with "communicated diseases" with nocebo effects likely to play an important role in the etiology of complaints rather than direct effects from turbines. This novel work highlighted the role of negative expectations and how they could lead to the development of complaints near wind farms. These findings were supported by many other studies that were suggestive of subjective variables, rather than wind turbine specific variables, as the source of annoyance for some people.

Whitfield Aslund et al. (36): Whitfield Aslund et al. used previously reported dose–response relationships between wind turbine noise and annoyance to predict the level of community noise annoyance that may occur in the province of Ontario. Prediction for future wind farm developments (planned, approved, or in process) were compared to previously reported rates of annoyance that were associated with more common noise sources (e.g., road traffic). Modeled noise levels and distance to the nearest wind farm-related noise source were compiled for over 8,000 individual receptor locations (i.e., buildings, dwellings, campsites, places of worship, institutions, and/or vacant lots) from 13 wind power projects in the province of Ontario that had been approved since 2009 or were under Ministry of the Environment (MOE) review as of July 2012. This information was then compared to the wind turbine noise specific dose–response relationships for self-reported annoyance from Pedersen et al. (17) and Bakker et al. (26) using data collected from 725 survey respondents living in the proximity of wind turbines (<2.5 km) in the Netherlands.

One of the study findings was that a distinct exponentially decreasing relationship was observed between distance to the nearest noise source and the sound pressure level predicted. However, although distance to the nearest noise source could explain a large proportion (86%) of the total variance in predicted sound pressure levels, other sources of variation are also important; predicted sound pressure levels at a set distance varied by approximately 5–10 dB(A) and the distance at which a set sound pressure level was met varied by approximately 1000 m. These variations reflect differences in the noise model inputs such as the physical design and noise emission ratings of the turbines (and transformer substations, if present) used in different projects and the total number of turbines (and transformer substations, if present) in the vicinity of the receptor location. Given that noise levels can vary substantially at a given distance, these data highlighted the inadequacy of using distance to the nearest turbine as a proxy for wind turbine noise exposure.

One of the other findings was that, for non-participating receptors, predicted rates of noise-related annoyance (when indoors) would not exceed

8%, with further reductions in the rates of annoyance at increased distances (i.e., >1 km). In comparison, it had previously been established that approximately 8% of adult Canadians reported being either "very or extremely bothered, disturbed, or annoyed" by noise in general when they were at home and 6.7% of adult Canadians indicated they were either "very or extremely annoyed" by traffic noise specifically (54). Even in small Canadian communities (i.e., <5000 residents) that are typically associated with low background noise levels, 11% of respondents were moderately to extremely annoyed by traffic noise (54). This analysis suggested that the current wind turbine noise restrictions in Ontario will limit community exposure to wind turbine related noise such that levels of annoyance are unlikely to exceed previously established background levels of noise-related annoyance from other common noise sources.

### 10.3.3 LOW-FREQUENCY NOISE AND INFRASOUND

As reviewed by Knopper and Ollson (9), a number of sources have proposed that the self-reported health effects of some people living near wind turbines may be due to LFN and infrasound [e.g., (20, 39, 55)]. However, infrasound and LFN are not unique to wind turbines; natural sources of infrasound include meteors, volcanic eruptions, ocean waves, wind, and any effect that leads to slow oscillations of the air (11). Measured LFN and infrasound levels from wind turbines have been shown to comply with available standards and criteria published by numerous government agencies including the UK Department for Environment, Food, and Rural Affairs; the American National Standards Institute; and the Japan Ministry of Environment (22). Therefore, Knopper and Ollson (9) concluded that the hypothesis that infrasound is a causative agent in health effects does not appear to be supported. With some exceptions, more recent studies (summarized below) generally support this hypothesis.

Møller and Pedersen (37): Møller and Pedersen conducted a LFN study from four large turbines (>2 MW) and 44 other small and large turbines that were aggregated (7 > 2 and 37 < 2 MW). Low-frequency sound (LFS) insulation was measured for 10 rooms under normal living conditions in houses exposed to LFN. They concluded that the spectrum of wind turbine noise moves down in frequency with increasing turbine size. They also suggested that the low-frequency part of the noise spectrum plays an important role in the noise at neighboring properties. They hypothesized that if the noise from the investigated large turbines had an outdoor level of 44 dB(A) (the maximum of the Danish regulation for wind turbines) there was a risk that a substantial proportion of the residents would be annoyed by LFN, even indoors. However, the authors' work did not include a survey of annoyance surrounding the turbines and did not provide any data to support this hypothesis. In terms of infrasound (sound below 20 Hz), they concluded that the levels were relatively low when human sensitivity to these frequencies was accounted for. Even in close proximity to turbines, the infrasonic sound pressure level was below the normal hearing threshold. Overall, this study suggested that LFN could be an important component of the overall noise levels from wind turbines. However, it did not provide a link between modeled or measured values and potential health effects of nearby residents. Rather, it hypothesized that at 44 dB(A), at least a portion of the annoyance could be attributed to LFN levels.

Bolin et al. (38): Bolin et al. (38) conducted a literature review over a 6-month period ending April 2011 into the potential health effects related to infrasound and LFN exposure surrounding wind turbines. They conducted the search using PubMed, PsycInfo, and Science Citation Index. In addition, they conducted gray literature searches and personally contacted researchers and noise consultants working with wind turbine noise. They concluded that the dominant source of wind turbine generated LFN was from incoming turbulence interacting with the blades. They found no evidence in the literature that infrasound in the 1–20 Hz range contributed to perceived annoyance or other health effects. They also opined that LFN from modern wind turbines could be audible at typical levels in residential settings, but did not exceed levels from other common noise sources, such as road traffic noise.

The authors concluded that empirical support was lacking for claims that LFN and infrasound cause serious health affects in the form of "vibroacoustic disease (VAD)," "wind turbine syndrome," or harmful effects on the inner ear. This conclusion was similar to that provided in the Massachusetts Department of Environmental Protection (MassDEP) and

Massachusetts Department of Public Health (MDPH) expert panel review released in January 2012.

Rand et al. (39) and Ambrose et al. (40): in the fall of 2011, Rand et al. published their findings on noise measurements taken around a residential home online in the Bulletin of Science, Technology and Society (BSTS) (39). In 2012, a similar article appeared in BSTS, but with Ambrose as first author. After learning about reported noise and health issues from some residents living near three wind turbines (Vestas, Model V82, 1.65 MW each) in Falmouth, MA, USA, Ambrose et al. conducted a study to investigate the role of infrasound and LFS in these complaints. What led Ambrose et al. to focus on infrasound and LFS was the home owner's complaints about discomfort and a number of symptoms (i.e., headaches, ear pressure, dizziness, nausea, apprehension, confusion, mental fatigue, inability to concentrate, and lethargy). These observations were reported to be associated with being indoors when the wind turbines were operating during moderate to strong winds. Ambrose et al. state: "Typically, indoors the A-weighted sound level is lower than outdoors when human activity is at a minimum. This strongly suggested that the A-weighted sound level might not correlate very well [sic] the wind turbine complaints. This may be indicative of another cause such as low- or very-low-frequency energy being involved."

The authors made acoustic measurements and viewed the data with dBL (unweighted) and dB(A), (C), and (G) filtering between April 17 and 19, 2011, at four locations [260 ft (~87 m), 830 ft (~277 m), 1,340 ft (~450 m), and 1,700 ft (~570 m)] between one turbine and one residence. The relationship between sound [dB(A), (G), and (L)] and health effects was based on measurements at 1,700 ft. Ambrose et al. reported that within 20 min, both authors had difficulties performing ordinary tasks and within 1 h both were "debilitated and had to work much harder mentally." They also claimed that as time went on their symptoms became more severe.

The authors reported being affected when wind speeds were greater than 10 m/s at the hub height of the turbines and when measured sound levels were in the 18–24 dB(A) range inside [51–64 dB(G); 62–74 dB(L)] and 32–46 dB(A) outside [49–65 dB(G); 57–69 dB(L)]. They reported that they felt effects inside and outside but preferred being outside. They noted that it took a week to recover but one researcher had recurring symptoms (of nausea and vertigo) for over 7 weeks. There are a number of uncertainties in the Ambrose et al. white paper and the BSTS articles, which diminished the strength of their conclusions. This was the first written account we are aware of that suggested acute health effects from exposure to sound from wind turbines. The recent MassDEP and MDPH (56) report provided this comment regarding the Ambrose et al. study: "Importantly, while there is an amplification at these lower frequencies, the indoor levels (unweighted) are still far lower than any levels that have ever been shown to cause a physical response (including the activation of the OHC) in humans."

Further, studies where biological effects observed following infra-sound exposure were conducted at sound pressure levels much greater than measured by Ambrose et al. [e.g., (11); 145 and 165 dB; (57): 130 dB] and much greater than what is produced by wind turbines. There are over 100,000 wind turbines in operation globally. Indeed, the idea of overt acute debilitating effects (even lasting several weeks after removal from exposure) appears to be unique to these authors.

Turnbull et al. (41): Turnbull et al. developed an underground tech-nique to measure infrasound and applied this process at two Australian wind farms as well as in the vicinities of a beach, a coastal cliff, the city of Adelaide, and a power station. The measured levels were compared against one another and against the infrasound audibility threshold of 85 dB(G). The authors reported that the measured level of infrasound within the wind farms was well below the audibility threshold and was similar to that of urban and coastal environments and near other engineered noise sources. Indeed, the level of infrasound from wind farms at 360 and 85 m [61 and 72 dB(G), respectively] was comparable to that observed at a distance of 25 m from ocean waves [75 dB(G)].

Crichton et al. (7): this study examined the possibility that expectations of negative health effects from exposure to infrasound promote symptom reporting. A sham controlled, double-blind provocation study was con-ducted in which participants were exposed to 10 min of infrasound and 10 min of sham infrasound. A total of 54 participants (34 women, 20 men) were randomized into high- or low-expectancy groups and presented with audiovisual information (including internet material) designed to invoke either high or low expectations that exposure to infrasound causes specific symptoms (e.g., headache, ear pressure, itchy skin, sinus pressure, dizziness, vibrations within the body). Notably, participants in the high-expectancy group reported significant increases in the number and intensity of symptoms experienced during exposure to both infrasound and sham infrasound. Conversely, there were no symptomatic changes in the low-expectancy group.

Based on their findings, Crichton et al. (7) concluded: "Healthy volunteers, when given information about the expected physiological effect of infrasound, reported symptoms that aligned with that information, during exposure to both infrasound and sham infrasound. Symptom expectations were created by viewing information readily available on the Internet, indicating the potential for symptom expectations to be created outside of the laboratory, in real world settings. Results suggest psychological expectations could explain the link between wind turbine exposure and health complaints." These results were consistent with the findings of other researchers, who have observed increased concern about the health risks associated with exposure to certain environmental hazards can lead to elevated symptom reporting, even when no objective health risk is presented (58, 59).

Crichton et al. (8): building on their previous publication that negative expectations established by the media and internet can significantly increase health-related complaints by exposed individuals (8), the authors investigated how positive expectations can produce a reduction in symptoms. Sixty participants were exposed to audible wind farm sound [43 dB(A)] and infrasound [9 Hz, 50.4 dBL (unweighted)] previously recorded 1 km from a wind farm, in two, 7 min session. Following baseline measurements, expectations were developed by watching videos that either promoted the negative health effects or the potentially therapeutic health effects of exposure to infrasound. Expectations were found to significantly alter symptom reporting: participants who were primed with negative expectations became more symptomatic over time, suggesting that their experiences during the first exposure session reinforced expectations and led to heightened symptomatic experiences in subsequent sessions. Upwards of 77% of participants in the negative expectation group reported a worsening of symptoms. In contrast, 90% of participants in the positive expectation group reported improvements in physical symptoms after the listening session. This was the first study to show that a placebo response could be brought on by positive pre-exposure expectations and influence participants exposed to wind farm noise. The authors concluded that negative expectations created by the media could account for the increase in negative health effects reported by individuals exposed to wind farm noise. Overall, this investigation provided further evidence that physiological outcomes can be influenced by established expectations.

### 10.3.4 ELECTROMAGNETIC FIELDS

Concerns about the ever-present nature of EMF (also called electric and magnetic fields) and possible health effects have been raised by some in the global community for a number of years. However, the science around EMF and possible health concerns has been extensively researched, with tens of thousands of scientific studies published on the issue. Government and medical agencies including Health Canada (60), the World Health Organization (61), the International Commission on Non-Ionizing Radiation Protection (62), the International Agency for Research on Cancer (63), and the US National Institute of Health (NIH) and National Institute of Environmental Health Sciences (64) have all thoroughly reviewed the available information. While individual opinions on the issue vary, the weight of scientific evidence does not support a causal link between EMF and health issues at levels typically encountered by people.

Short-term exposure to EMF at high levels is known to cause nerve and muscle stimulation in the central nervous system. Based on this information, the ICNIRP, a group recognized by the WHO as the international independent advisory body for non-ionizing radiation protection, established an acute exposure guideline of 2,000 mG for the general public, based on power frequency EMF of 50–400 Hz (62). With respect to long-term exposure to low levels of EMF, it needs to be acknowledged that the IARC and WHO have categorized EMF as a Class 2B possible human carcinogen, based on a weak association of childhood leukemia and magnetic field strength above 3–4 mG (63). This means there is limited evidence of carcinogenicity in humans and inadequate evidence of carcinogenicity in experimental animals. These human studies are weakened by various methodological problems that the WHO has identified as a combination of selection bias, some degree of confounding and chance (65). There are also no globally accepted mechanisms that would suggest that low-level exposures are involved in cancer development and animal studies have been largely negative (65). Thus, the WHO has stated that, based on approximately 25,000 articles published over the past 30 years, the evidence linking childhood leukemia to EMF exposure is not strong enough to be considered causal (61). Concerns have also been raised by some about a relationship between EMF and a range of various health concerns, including cancers in adults, depression, suicide, and reproductive dysfunction, among several others. The WHO (65) has stated: "...scientific evidence supporting an association between ELF [extremely low frequency] magnetic field exposure and all of these health effects is much weaker than for childhood leukaemia."

Recently, worries about exposure to EMF from wind turbines, and associated electrical transmission, has been raised at public meetings and legal proceedings. These fears have not been based on any actual measurements of EMF exposure surrounding existing projects but appear to follow from concerns raised from internet sources and misunderstanding of the science. There has been limited research conducted on wind turbine emissions of EMF, either from the turbines themselves, or from the power lines required for distribution of the generated electricity. However, based on the weight of evidence it is not expected that EMF from wind turbines is likely to be a causative agent for negative health effects in the community. Only three papers were retrieved in the preparation of this review that examined this issue specifically.

Havas and Colling (42): the paper indicated that there were some people who lived around wind turbines that complained of difficulty sleeping, fatigue, depression, irritability, aggressiveness, cognitive dysfunction, chest pain/pressure, headaches, joint pain, skin irritations, nausea, dizziness, tinnitus, and stress. The authors suggested that these symptoms could be caused by electromagnetic waves in the form of poor power quality (dirty electricity) and ground current resulting in health effects in those that are electrically hypersensitive. They indicated that individuals reacted differently to both sound and electromagnetic waves and this could explain why not everyone experienced the same health effects living near turbines. Ground current or stray voltage was also purported to be a potential cause of health effects surrounding wind turbines. However, this paper was hypothetical and speculative in nature and no data were presented to support the author's opinions. Presently, there are no quantitative data in the scientific literature to support the claims made in Havas and Colling (42).

Israel et al. (43): these authors conducted EMF, sound, and vibration measurements surrounding one of the largest wind energy parks in Bulgaria, located along the Black Sea. The purpose of the study was to determine if levels of wind turbine emissions were within Bulgarian and European limits for workers and the general population. In addition, they sought to determine if their previously established 500 m setback zone around the wind park was adequate. The wind park consisted of 55 Vestas V90 3 MW towers. The measurements took place over a 72-h period when temperatures were between 0 and 5.5°C. Actual distances to the receptor locations were not reported, although it is suspected that they would be in the vicinity of 500 m from the closest turbines.

The EMF levels measured within 2–3 m of the wind turbines were between 0.133 and 0.225 mG. These values are comparable to or lower than magnetic field measurements that have been reported in the proximity of typical household electrical devices (66). It should be noted that the values observed by Israel et al. were approximately four orders of magnitude lower than the ICNIRP (62) guideline of 2,000 mG for the general public for acute exposure. Based on these findings, Israel et al. concluded that the EMF levels from wind turbines were at such low level as to be insignificant compared to values found in residential areas and homes. The findings reported by Israel et al. of actual measurements of EMF surrounding wind turbines were contrary to the hypothesis presented by Havas and Colling (42).

The noise measurements performed by Israel et al. met the requirements of Bulgarian legislation for day [55 dB(A)], evening [50 dB(A)], and night [45 dB(A)] and it was concluded that the wind turbines contributed only 1–3 dB(A) above existing background levels. Vibration measurements surrounding the turbines had values close to zero, which indicated that this was not a contributing emission factor of exposure for people living around wind turbines. Overall, the authors concluded:"…the studied wind power park complies with the requirements of the national and European legislation for human protection from physical factors–electric and magnetic fields up to 1 kHz, noise, vibration, and do not create risk for both workers in the area of the park and the general population living in the nearest villages."

McCallum et al. (44): this study was carried out at the Kingsbridge 1 Wind Farm located near Goderich, ON, Canada. Magnetic field measurements (milligauss) were collected in the proximity of 15 Vestas 1.8 MW wind turbines, two substations, various buried and overhead collector and transmission lines, and nearby homes. Data were collected during three operational scenarios to characterize potential EMF exposure: "high wind" (generating power), "low wind" (drawing power from the grid, but not generating power), and "shut off" (neither drawing, nor generating power).

Background levels of EMF (0.2–0.3 mG) were established by measuring magnetic fields around the wind turbines under the "shut off" scenario. Magnetic field levels detected at the base of the turbines under both the "high wind" and "low wind" conditions were low (mean = 0.9 mG; n = 11) and rapidly diminished with distance, becoming indistinguishable from background within 2 m of the base. Magnetic fields measured 1 m above buried collector lines were also within background (≤0.3 mG). Beneath overhead 27.5 and 500 kV transmission lines, magnetic field levels of up to 16.5 and 46 mG, respectively, were recorded. These levels also diminished rapidly with distance. None of these sources appeared to influence magnetic field levels at nearby homes located as close as just over 500 m from turbines, where measurements immediately outside of the homes were ≤0.4 mG. The results suggested that there was nothing unique to wind farms with respect to EMF exposure; in fact, magnetic field levels in the vicinity of wind turbines were lower than those produced by many common household electrical devices (e.g., refrigerator, dishwasher, microwave, hairdryer) and were well below any existing regulatory guidelines with respect to human health.

### 10.3.5 SHADOW FLICKER

The main health concern associated with shadow flicker is the risk of seizures in those people with photosensitive epilepsy. As reviewed by

Knopper and Ollson (9), Harding et al. (14) and Smedley et al. (19) have published the seminal studies dealing with this concern. Both authors investigated the relationship between photo-induced seizures (i.e., photosensitive epilepsy) and wind turbine blade flicker (also known as shadow flicker). Both studies suggested that flicker from turbines that interrupt or reflect sunlight at frequencies >3 Hz pose a potential risk of inducing photosensitive seizures in 1.7 people per 100,000 of the photosensitive population. For turbines with three blades, this translates to a maximum speed of rotation of 60 rpm. Modern turbines commonly spin at rates well below this threshold. For example, the following spin rates for four different models of wind turbines have been obtained from the turbine specification sheets:

- Siemens SWT-2.3: 6–16 rpm
- REpower MM92: 7.8–15.0 rpm
- GE 1.6–100: 9.75–16.2 rpm
- Vestas V112-3.0: 6.2–17.1 rpm

In 2011, the Department of Energy and Climate Change (67) released a consultant's report entitled "Update of UK Shadow Flicker Evidence Base." The report concluded that: "On health effects and nuisance of the shadow flicker effect, it is considered that the frequency of the flickering caused by the wind turbine rotation is such that it should not cause a significant risk to health." Furthermore, the expert panel convened by MassDEP and MDPH (56) concluded that the scientific evidence suggests that shadow flicker does not pose a risk of inducing seizures in people with photosensitive epilepsy.

Germany is one of the only countries to implement formal shadow flicker guidelines, which are part of the Federal Emission Control Act (68). These guidelines allow:

- maximum 30 h per year of astronomical maximum shadow (worst case);
- maximum 30 min worst day of astronomical maximum shadow (worst case); and
- maximum 8 h per year actual.

Although shadow flicker from wind turbines is unlikely to lead to a risk of photo-induced epilepsy, there has been little if any research conducted on how it could heighten the annoyance factor of those living in proximity to turbines. It may however be included in the notion of visual cues.

### 10.3.6 REVIEW ARTICLES, EDITORIALS, AND SOCIAL COMMENTARIES

In addition to the articles reviewed above that reported the results of surveys and experiments designed to specifically investigate potential environmental stressors that have been associated with wind turbines (i.e., overall noise, LFN and infrasound, EMF, and shadow flicker), a number of published and peer-reviewed articles were identified that present reviews of the available data, opinion pieces, and/or social commentaries. These articles are reviewed in detail below.

Bulletin of Science, Technology and Society: Special Edition 2011, 31(4): in August 2011, authors of a number of popular literature studies published their findings as a series of nine articles in a special edition of the Bulletin of Science, Technology and Society (BSTS) devoted entirely to wind farms and potential health effects[1]. Many of the articles in the special edition were written as opinion pieces or social commentaries and did not provide detailed methodologies used to test hypotheses as is expected in the publication of scientific research articles. Based on a critical review of each of the articles (69), it is our opinion that the series suffers numerous flaws from a scientific, technological, and social basis. Many of the claims used as evidence of a relationship between health effects and wind turbines were unsubstantiated [e.g., Phillips (70) is entirely unsupported and contains alarmist extrapolations], without proper references [e.g., (70, 71)] and based on anecdotal or unconfirmed reports [e.g., (55, 70, 72, 73)], fallacious comparisons [e.g., (74)], and reaching arguments lacking a logical process [e.g., (70, 73, 75, 76)]. Further, much information given as fact was contrary to that published in the scientific literature; indeed, many authors appeared to selectively reference articles and information in a way that would benefit their own arguments [e.g., (55, 71)]. The results of this BSTS special issue failed to provide valid, defensible scientific and social arguments to suggest that wind turbines, regardless of siting considerations, cause harm to human health.

Hanning and Evans (45) and Chapman (46): in 2012, Hanning and Evans had an editorial published in the British Medical Journal (BMJ), the purpose of which was to opine on the relationship between wind turbines noise and health effects. By citing a short list of articles (12), half of which are from the non-indexed journal BSTS or from conference proceedings (3 and 3, respectively, out of 12), Hanning and Evans suggested that: "A large body of evidence now exists to suggest that wind turbines disturb sleep and impair health at distances and external noise levels that are permitted in most jurisdictions." and "Robust independent research into the health effects of existing wind farms is long overdue, as is an independent review of existing evidence and guidance on acceptable noise levels."

Shortly after publication, this editorial was rebuffed by Chapman (46), in another editorial placed in the BMJ. Chapman pointed out that there are a number of independent reviews of the literature around wind turbines and human health (Chapman points to 17 such papers not referenced by Hanning and Evans). Chapman opined that: "These reviews strongly state that the evidence that wind turbines themselves cause problems is poor. They conclude that: Small minorities of exposed people claim to be adversely affected by turbines; Negative attitudes to turbines are more predictive of reported adverse health effects and annoyance than are objective measures of exposure; Deriving income from hosting wind turbines may have a "protective effect" against annoyance and health symptoms." Further debate about the original editorial is available online to view (and comment on) through the BMJ web site[2].

Farboud et al. (47): this review article looked at the effects of LFN and infrasound and questioned the existence of "wind turbine syndrome." The authors conducted a literature search for articles published within the last 10 years, using the PubMed database and the Google Scholar search engine. Their search terms included "wind turbine," "infrasound," or "LFN" and search results were limited to the English language, human trials, and either randomized control trials, meta-analyses, editorial letters, clinical trials, case reports, comments, or journal articles. A number of articles dealing with "wind turbine," "infrasound," or "LFN," and available in PubMed and Google Scholar, appear to have been missed by Farboud et al. [e.g., (9, 22, 38)]. The review included discussions on topics such as wind turbine noise measurements and regulations, wind turbine syndrome, and the effects of LFN and infrasound.

The authors discussed the use of A-weighting in noise measurements from wind turbines stating: "The A-filter de-emphasizes all auditory energy with frequencies of less than 500 Hz, and completely ignores all auditory energy of less than 20 Hz, in an effort to estimate the noise thought to be actually processed by the ear. Hence, much of the noise produced by a wind turbine is effectively ignored." The authors later described the results and implications of studies looking at the effects of infrasound in the ear, and noted that infrasound and LFN are currently not recognized as disease agents. Referencing a study by Salt and Hullar (20), the authors noted that the inner hair cells of the cochlea, which is the main hearing pathway in mammals, are not sensitive to infrasound. Conversely, the outer hair cells of the cochlea are more sensitive to LFN and infrasound and can be stimulated at levels below the auditory threshold. Nevertheless, the authors conceded that: "...low-frequency noise may well influence inner ear physiology. However, whether this actually alters function or causes symptoms is unknown."

It should be noted that, as discussed in the "Low-Frequency Noise and Infrasound" section of this review, there were a number of studies that specifically addressed the concerns of LFN and infrasound from wind turbines that suggested that these were unlikely to be causative agents in health effects of those living near wind turbines [e.g., (7, 11, 22, 37, 38)]. Unfortunately, none of these studies were included as part of the Farboud et al. review.

Regarding the existence of "Wind Turbine Syndrome," Farboud et al. stated that: "There is an abundance of information available on the internet describing the possibility of wind turbine syndrome. However, the majority of this information is based on purely anecdotal evidence." The authors briefly discussed the various symptoms that have been self-reported by individuals and attributed to noise from wind turbines. They also pointed out that "Wind Turbine Syndrome" was not a clinically recognized diagnosis, remained unproven, and was not generally accepted within the scientific and medical community. They also mentioned that some researchers maintained that the effects of "Wind Turbine Syndrome" were just examples of the well-known stress effects of exposure to noise, as displayed by a small proportion of the population.

Farboud et al. concluded their review by suggesting that the evidence available was incomplete and until the physiological effects of LFN and infrasound were fully understood, it was not possible to conclusively state that wind turbines were not causing any of the reported effects. However, it was not clear how this conclusion might have been altered had they considered the additional available information regarding LFN and infrasound from wind turbines described elsewhere in this review [i.e., (7, 11, 22, 37, 38)].

McCubbin and Sovacool (48): McCubbin and Sovacool (48) presented a comparison of the health and environmental benefits of wind power in contrast to natural gas. The authors selected two locations: the 580 MW wind farm at Altamont Pass in California and the 22 MW wind farm in Sawtooth, ID, USA. The paper considered the environmental and economic benefits associated with each wind farm. Human health benefits were calculated based on a reduction in ambient PM2.5 levels using well-established health impact and valuation functions from the US EPA. Additionally, benefits to the health and well-being of wildlife and avian species were quantified.

With regard to the human health impacts, the potential cost savings were associated with effects such as premature mortality, hospital admissions, emergency rooms visits, asthma attacks, and respiratory symptoms. The details of the quantification methods and equations used to calculate the benefits to externalities such as human health, wildlife, and the natural environment were not provided herein but are available in the published manuscript.

McCubbin and Sovacool determined that from 2012 to 2031 the wind turbines at Altamont Pass will avoid anywhere from $560 million to $4.38 billion in human health and climate-related externalities, and the Sawtooth wind farm will avoid from $18 million to $24 million. The authors noted that there were uncertainties associated with their quantification methods and final cost estimates; however, they claimed that the values were likely underestimated based on numerous factors that were not considered (e.g., other pollutants). They concluded that: "Despite the uncertainties, the evidence gathered here strongly suggests that natural gas had substantial external costs that should be included in an evaluation comparing wind energy to combined cycle natural gas-fired power plants. The overall costs of electricity generated by natural gas are greater than those from wind energy when environmental and human health externalities are quantified. It remains likely that over time the relative difference will widen, making the use of wind energy even more favorable."

Roberts and Roberts (49): the authors conducted a summary of the peer-reviewed literature on the research that examined the relationship between human health effects and exposure to LFS and sound generated from the operation of wind turbines. The PubMed database (maintained by the US National Library of Medicine) was relied upon for retrieving the peer-reviewed literature used in this review. A number of search terms were used including: "infrasound and health effects"; "LFN and health effects"; "LFS and health effects"; "wind power and noise"; and "wind turbines AND noise." In total, 156 articles were identified with 28 articles addressing health effects and LFS related to wind turbines. Based on the collective results of the studies reviewed, Roberts and Roberts (49) found that: "At present, a specific health condition or collection of symptoms has not been documented in the peer-reviewed, published literature that has been classified as a 'disease' caused by exposure to sound levels and frequencies generated by the operations of wind turbines. It can be theorized that reported health effects are a manifestation of the annoyance that individuals experience as a result of the presence of wind turbines in their communities."

Chapman and St. George (50): in 2007, Alves-Pereira and Castelo Branco issued a press-release suggesting that their research demonstrated that living in proximity to wind turbines had led to the development of VAD in nearby home-dwellers (9). Alves-Pereira and Castelo Branco appear to be the primary researchers who have circulated VAD as a hypothesis for adverse health effects and wind turbines and to our knowledge this work has never appeared in a peer-reviewed article. In this paper, Chapman and St. George investigated the extent to which VAD and its alleged association with wind turbine exposure had received scientific attention, the quality of that association, and how the alleged association gained support by wind farms opponent.

Based on a structured scientific database and Google search strategy, the authors showed that "VAD has received virtually no scientific recognition beyond the group who coined and promoted the concept. There is no evidence of even rudimentary quality that vibroacoustic disease is associated with or caused by wind turbines." They went on to state that an implication of this "factoid"—defined as questionable or spurious statements—may have been contributing to nocebo effects among those living near turbines. That is the spread of negative, often emotive information would be followed by increases in complaints and that without such suggestions being spread, complaints would be less. These results highlighted the role that perception plays in the human health wind turbine debate and underscored the role of proper risk communication in communities.

Jeffery et al. (51, 52): the overall goal of these commentary pieces was to provide information to physicians regarding the possible health effects of exposure to noise produced by wind turbines and how these may manifest in patients. In the 2013 article, information about the Green Energy Act was presented in such a way that implied that the overall goal of the Act was to remove protective noise regulations and allow wind turbines to be placed "in close proximity to family homes." The authors suggested that there has been a concerted effort to minimize the potential health risks while convincing the general public and physicians that wind turbines are beneficial. No evidence was given to support these claims. Case reports and publications that reported adverse effects following wind turbines noise exposure were briefly discussed; however, only the negative health effects were highlighted. Older literature and a number of non-peer-reviewed articles and media reports were used to support the author's opinions. The 2014 paper is very similar to that published in 2013. The authors provided a very one-sided opinion in their review of the issue of wind turbines and adverse health effects. They have missed a number of key and pertinent articles that have been published on the issue. Overall the authors did not provide adequate data or support for their arguments, in both papers, nor did they provide accurate information regarding the weight of scientific data on the issue.

### 10.3.7 WEIGHT OF EVIDENCE CONCLUSIONS

There are roughly 60 studies that have been conducted worldwide on the issue of wind turbines and human health. In terms of effects being related to wind turbine operational effects and wind turbine noise, there are fewer than 20 articles. The vast majority has been published in one journal (BSTS) and many of these authors sit on advisory board of the Society for Wind Vigilance, an advocacy group in the province of Ontario. However, with respect to effects being more likely attributable to a number of subjective variables (when turbines are sited correctly), there are closer to 45 articles. These articles are published by a variety of different authors with wide and diverse affiliations. Indeed, conclusions stemming from these articles are supported by studies where audible and inaudible noise has been quantified from operational wind turbines.

Based on the findings and scientific merit of the research conducted to date, it is our opinion that the weight of evidence suggests that when sited properly, wind turbines are not related to adverse health effects. This claim is supported (and made) by findings from a number of government health and medical agencies and legal decisions [e.g., (56, 77–80)]. Collectively, the evidence has shown that while noise from wind turbines is not loud enough to cause hearing impairment and is not causally related to adverse effects, wind turbine noise can be a source of annoyance for some people and that annoyance may be associated with certain reported health effects (e.g., sleep disturbance), especially at sound pressure levels >40 dB(A).

The reported correlation between wind turbine noise and annoyance is not unexpected as noise-related annoyance [described by Berglund and Lindvall (81) as a "feeling of displeasure evoked by a noise"] has been extensively linked to a variety of common noise sources such as rail, road, and air traffic (81–83). Noise-related annoyance from these more common sources is prevalent in many communities. For instance, results of national surveys in Canada and the U.K. by Michaud et al. (54) and Grimwood et al. (84), respectively, suggested that annoyance from noise (predominantly traffic noise) may impact approximately 8% of the general population. Even in small communities in Canada (i.e., <5000 residents) where traffic is relatively light compared to urban centers, Michaud et al. (54) reported that 11% of respondents were moderately to extremely annoyed by traffic noise.

Although annoyance is considered to be the least severe potential impact of community noise exposure (83, 85), it has been hypothesized that sufficiently high levels of annoyance could lead to negative emotional responses (e.g., anger, disappointment, depression, or anxiety) and psychosocial symptoms (e.g., tiredness, stomach discomfort, and stress) (83, 86–90). However, it is important to note that noise annoyance is known to be strongly affected by attitudinal factors such as fear of harm connected with the source and personal evaluation of the source (91–93) as well as expectations of residents (92). For wind turbines, this has been reflected in studies that have shown that subjective variables like evaluations of visual impact (e.g., beautiful vs. ugly), attitude to wind turbines (benign vs. intruders), and personality traits are more strongly related to annoyance and health effects than noise itself [e.g., (4, 5, 16, 17, 31)]. Thus, it is likely that the adverse effects exhibited by some people who live near wind turbines are a response to stress and annoyance, which are driven by multiple environmental and personal factors, and are not specifically caused by any unique characteristic of wind turbines. This hypothesis is also supported by the observation that people who economically benefit from wind turbines have significantly decreased levels of annoyance compared to individuals that received no economic benefit, despite exposure to similar, if not higher, sound levels (17).

There is also a growing body of research that suggests that nocebo effects may play a role in a number of self-reported health impacts related to the presence of wind turbines. Negative attitudes and worries of individuals about perceived environmental risks have been shown to be associated with adverse health-related symptoms such as headache, nausea, dizziness, agitation, and depression, even in the absence of an identifiable cause (94–96). Psychogenic factors, such as the circulation of negative information and priming of expectations have been shown to impact self-assessments following exposure to wind turbine noise (6–8). It is therefore important to consider the role of mass media in influencing public attitudes about wind turbines and how this may alter responses and perceived health impacts of wind turbines in the community. For example, Deignan et al. (97) recently demonstrated that newspaper coverage of the potential health effects of wind turbines in Ontario has tended to emphasize "fright factors" about wind turbines. Specifically, Deignan et al. (97) reported that 94% of articles provided "negative, loaded or fear-evoking" descriptions of "health-related signs, symptoms or adverse effects of wind turbine exposure" and 58% of articles suggested that the effects of wind turbines on human health were "poorly understood by science." It is possible that this type of coverage may have a significant impact on attitudinal factors, such as fear of the noise source, that are known to increase noise annoyance (91–93).

Stress/annoyance is not unique to living in proximity to wind turbines. The American Psychological Association (98) published a report stating that the majority of Americans are living with moderate (4 to 7 on a scale of 1 to 10) or high (8 to 10 on a scale of 1 to 10) levels of stress. APA identified money, work, and the economy as the most often cited sources of stress in Americans followed by family responsibilities, relationships, job stability, housing costs, health concerns, health problems, and safety. Stress from these and other sources can lead to a number of adverse health effects that are commonplace in society. The Mayo Clinic (99) identifies irritability, anger, anxiety, sadness/guilt, change in sleep, fatigue, difficulty concentrating or making decisions, loss of interest/enjoyment, nausea, headache, and tinnitus as common symptoms of stress. Interestingly, these symptoms are nearly identical to those suggested by McMurtry (55) as criteria for a "diagnosis of adverse health effects in the environs of industrial wind turbines."

Based on the available evidence, we suggest the following best practices for wind turbine development in the context of human health. However, it should be noted that subjective variables (e.g., attitudes and expectations) are strongly linked to annoyance and have the potential to facilitate other health complaints via the nocebo effect. Therefore, it is possible that a segment of the population may remain annoyed (or report other health impacts) even when noise limits are enforced.

1. Setbacks should be sound-based rather than distance-based alone.
2. Preference should be given to sound emissions of ≤40 dB(A) for non-participating receptors, measured outside, at a dwelling, and not including ambient noise. This value is the same as the WHO (Europe) night noise guideline (100) and has been demonstrated to result in levels of wind turbine community annoyance similar to, or lower than, known background levels of noise-related annoyance from other common noise sources.
3. Post construction monitoring should be common place to ensure modeled sound levels are within required noise limits.

4.  If sound emissions from wind projects is in the 40–45 dB(A) range for non-participating receptors, we suggest community consultation and community support.

5.  Setbacks that permit sound levels >45 dB(A) (wind turbine noise only; not including ambient noise) for non-participating receptors directly outside a dwelling are not supported due to possible direct effects from audibility and possible levels of annoyance above background.

6.  When ambient noise is taken into account, wind turbine noise can be >45 dB(A), but a combined wind turbine-ambient noise should not exceed >55 dB(A) for non-participating and participating receptors. Our suggested upper limit is based on WHO (100) conclusions that noise above 55 dB(A) is "considered increasingly dangerous for public health," is when "adverse health effects occur frequently, a sizeable proportion of the population is highly annoyed and sleep-disturbed" and "cardiovascular effects become the major public health concern, which are likely to be less dependent on the nature of the noise."

Over the past 20 years, there has been substantial proliferation in the use of wind power, with a global increase of over 50-fold from 1996 to 2013 (1). Such an increase of investment in renewable energy is a critical step in reducing human dependency on fossil fuel resources. Wind-based energy represents a clean resource that does not produce any known chemical emissions or harmful wastes. As highlighted in a recent editorial in the British Medical Journal, reducing air pollution can provide significant health benefits, including reducing asthma, chronic obstructive pulmonary disease, cancer, and heart disease, which in turn could provide significant savings for health care systems (101). By following our proposed health-based best practices for wind turbine siting, wind energy developers, the media, members of the public and government agencies can work together to ensure that the full potential of this renewable energy source is met.

## FOOTNOTES

1.  http://bst.sagepub.com/
2.  http://www.bmj.com/content/344/bmj.e1527?tab=responses

# REFERENCES

1. GWEC (Global Wind Energy Council). Global Wind Energy Statistics 2013. (2014). Available from: http://www.gwec.net/wp-content/uploads/2014/02/GWEC-PRstats-2013_EN.pdf

2. Upham P, Whitmarsh L, Poortinga W, Purdam K, Darnton A, McLachlan C, et al. Public Attitudes to Environmental Change: A Selective Review of Theory and Practice. Swindon, UK: Economic and Social Research Council/Living with Environmental Change Programme (2009).

3. Pierpont N. Wind Turbine Syndrome. Santa Fe, NM: K-Selected Books (2009).

4. Pedersen E, Persson Waye K. Perception and annoyance due to wind turbine noise – a dose–response relationship. J Acoust Soc Am (2004) 116:3460–70. doi: 10.1121/1.1815091

5. Pedersen E, Persson Waye K. Wind turbine noise, annoyance and self-reported health and well-being in different living environments. Occup Environ Med (2007) 64:480–6. doi:10.1136/oem.2006.031039

6. Chapman S, St George A, Waller K, Cakic V. The pattern of complaints about Australian wind farms does not match the establishment and distribution of turbines: support for the psychogenic, 'communicated disease' hypothesis. PLoS One (2013) 8:e76584. doi:10.1371/journal.pone.0076584

7. Crichton F, Dodd G, Schmid G, Gamble G, Cundy T, Petrie KJ. Can expectations produce symptoms from infrasound associated with wind turbines? Health Psychol (2014) 33:360–4. doi:10.1037/a0031760

8. Crichton F, Dodd G, Schmid G, Gamble G, Cundy T, Petrie KJ. The power of positive and negative expectations to influence reported symptoms and mood during exposure to wind farm sound. Health Psychol (2013). doi:10.1037/hea0000037

9. Knopper LD, Ollson CA. Health effects and wind turbines: a review of the literature. Environ Health (2011) 10:78. doi:10.1186/1476-069X-10-78

10. van den Berg GP. Effects of the wind profile at night on wind turbine sound. J Sound Vib (2003) 277:955–70. doi:10.1016/j.jsv.2003.09.050

11. Leventhall G. Infrasound from wind turbines – fact, fiction or deception? Can Acoust (2006) 34:29–36.

12. Pedersen E, Hallberg LRM, Persson Waye K. Living in the vicinity of wind turbines – a grounded theory study. Qual Res Psychol (2007) 4:49–63. doi:10.1080/14780880701473409

13. Keith SE, Michaud DS, Bly SHP. A proposal for evaluating the potential health effects of wind turbine noise for projects under the Canadian Environmental Assessment Act. J Low Freq Noise Vib Active Control (2008) 27:253–65. doi:10.1260/026309208786926796

14. Harding G, Harding P, Wilkins A. Wind turbines, flicker, and photosensitive epilepsy: characterizing the flashing that may precipitate seizures and optimizing guidelines to prevent them. Epilepsia (2008) 49:1095–8. doi:10.1111/j.1528-1167.2008.01563.x

15. Pedersen E, Persson Waye K. Wind turbines – low level noise sources interfering with restoration? Environ Res Lett (2008) 3:1–5. doi:10.1088/1748-9326/3/1/015002

16. Pedersen E, Larsman P. The impact of visual factors on noise annoyance among people living in the vicinity of wind turbines. J Environ Psychol (2008) 28:379–89. doi:10.1016/j.scitotenv.2012.03.005

17. Pedersen E, van den Berg F, Bakker R, Bouma J. Response to noise from modern wind farms in The Netherlands. J Acoust Soc Am (2009) 126:634–43. doi:10.1121/1.3160293

18. Pedersen E, van den Berg F, Bakker R, Bouma J. Can road traffic mask the sound from wind turbines? Response to wind turbine sound at different levels of road traffic. Energ Policy (2010) 38:2520–7. doi:10.1016/j.enpol.2010.01.001

19. Smedley ARD, Webb AR, Wilkins AJ. Potential of wind turbines to elicit seizures under various meteorological conditions. Epilepsia (2010) 51:1146–51. doi:10.1111/j.1528-1167.2009.02402.x

20. Salt AN, Hullar TE. Responses of the ear to low frequency sounds, infrasound and wind turbines. Hear Res (2010) 268:12–21. doi:10.1016/j.heares.2010.06.007

21. Pedersen E. Health aspects associated with wind turbine noise – results from three field studies. Noise Control Eng J (2011) 59:47–53. doi:10.3397/1.3533898

22. O'Neal RD, Hellweg RD Jr, Lampeter RM. Low frequency noise and infrasound from wind turbines. Noise Control Eng J (2011) 59:135–57. doi:10.3397/1.3549200

23. Shepherd D, McBride D, Welch D, Dirks KN, Hill EM. Evaluating the impact of wind turbine noise on health related quality of life. Noise Health (2011) 13:333–9. doi:10.4103/1463-1741.85502

24. Janssen SA, Vos H, Pedersen E. A comparison between exposure-response relationships for wind turbine annoyance and annoyance due to other noise sources. J Acoust Soc Am (2011) 130:3746–53. doi:10.1121/1.3653984

25. Verheijen E, Jabben J, Schreurs E, Smith KB. Impact of wind turbine noise in The Netherlands. Noise Health (2011) 13:459–63. doi:10.4103/1463-1741.90331

26. Bakker RH, Pedersen E, van den Berg GP, Stewart RE, Lok W, Bouma J. Impact of wind turbine sound on annoyance, self-reported sleep disturbance and psychological distress. Sci Total Environ (2012) 425:42–51. doi:10.1016/j.scitotenv.2012.03.005

27. Nissenbaum MA, Aramini JJ, Hanning CD. Effects of industrial wind turbine noise on sleep and health. Noise Health (2012) 12:237–43. doi:10.4103/1463-1741.102961

28. Ollson CA, Knopper LD, McCallum LC, Whitfield-Aslund ML. Are the findings of "effects of industrial wind turbine noise on sleep and health" supported? Noise Health (2013) 15:68–71. doi:10.4103/1463-1741.110302

29. Barnard M. Letter to editor: issues of wind turbine noise. Noise Health (2013) 63:150–2. doi:10.4103/1463-1741.110305

30. Mroczek B, Kurpas D, Karakiewicz B. Influence of distances between places of residence and wind farms on the quality of life in nearby areas. Ann Agric Environ Med (2012) 19:692–6.

31. Taylor J, Eastwick C, Wilson R, Lawrence C. The influence of negative oriented personality traits on the effects of wind turbines noise. Pers Individ Diff (2012) 54:338–43. doi:10.1016/j.paid.2012.09.018

32. Evans T, Cooper J. Comparison of predicted and measured wind farm noise levels and implications for assessments of new wind farms. Acoust Aust (2012) 40:28–36.

33.  Maffei L, Iachini T, Masullo M, Aletta F, Sorrentino F, Senese VP, et al. The effects of vision-related aspects on noise perception of wind turbines in quiet areas. Int J Environ Res Public Health (2013) 10:1681–97. doi:10.3390/ijerph10051681

34.  Van Renterghem T, Bockstael A, De Weirt V, Bottledooren D. Annoyance, detection and recognition of wind turbine noise. Sci Total Environ (2013) 456:333–45. doi:10.1016/j.scitotenv.2013.03.095

35.  Baxter J, Morzaria R, Hirsch R. A case-control study of support/opposition to wind turbines: perceptions of health risk, economic benefit, and community conflict. Energy Policy (2013) 61:931–43. doi:10.1016/j.enpol.2013.06.050

36.  Whitfield Aslund ML, Ollson CA, Knopper LD. Projected contributions of future wind farm development to community noise and annoyance levels in Ontario, Canada. Energ Policy (2013) 62:44–50. doi:10.1016/j.enpol.2013.07.070

37.  Møller H, Pedersen CS. Low-frequency noise from large wind turbines. J Acoust Soc Am (2011) 129:3727–44. doi:10.1121/1.3543957

38.  Bolin K, Bluhm G, Eriksson G, Nilsson ME. Infrasound and low frequency noise from wind turbines: exposure and health effects. Environ Res Lett (2011) 6:106. doi:10.1088/1748-9326/6/3/035103

39.  Rand RW, Ambrose SE, Krogh CME. Occupational health and industrial wind turbines: a case study. Bull Sci Technol Soc (2011) 31:359–62. doi:10.1177/0270467611417849

40.  Ambrose SE, Rand RW, Krogh CME. Wind turbine acoustic investigation: infrasound and low-frequency noise: a case study. Bull Sci Technol Soc (2012) 32:128–41. doi:10.1177/0270467612455734

41.  Turnbull C, Turner J, Walsh D. Measurement and level of infrasound from wind farms and other sources. Acoust Aust (2012) 40:45–50.

42.  Havas M, Colling D. Wind turbines make waves: why some residents near wind turbines become ill. Bull Sci Technol Soc (2011) 31:414–26. doi:10.1177/0270467611417852

43.  Israel M, Ivanova P, Ivanova M. Electromagnetic fields and other physical factors around wind power generators (pilot study). Environmentalist (2011) 31:161–8. doi:10.1007/s10669-011-9315-z

44.  McCallum LC, Whitfield Aslund ML, Knopper LD, Ferguson GM, Ollson CA. Measuring electromagnetic fields (EMF) around wind turbines in Canada: is there a human health concern? Environ Health (2014) 13:9. doi:10.1186/1476-069X-13-9

45.  Hanning CD, Evans A. Wind turbine noise seems to affect health adversely and an independent review of evidence is needed. BMJ (2012) 344:e1527. doi:10.1136/bmj.e1527

46.  Chapman S. Editorial ignored 17 reviews on wind turbines and health. BMJ (2012) 344:e3366. doi:10.1136/bmj.e3366

47.  Farboud A, Crunkhorn R, Trinidade A. Wind turbine syndrome: fact or fiction? J Laryngol Otol (2013) 127:222–6. doi:10.1017/S0022215112002964

48.  McCubbin D, Sovacool BK. Quantifying the health and environmental benefits of wind power to natural gas. Energy Policy (2013) 53:429–41. doi:10.1016/j.enpol.2012.11.004

49.  Roberts JD, Roberts MA. Wind turbines: is there a human health risk? J Environ Health (2013) 75:8–17.

50. Chapman S, St George A. How the factoid of wind turbines causing 'vibroacoustic disease' came to be 'irrefutably demonstrated'. Aust N Z J Public Health (2013) 37:244–9. doi:10.1111/1753-6405.12066

51. Jeffery RD, Krogh C, Horner B. Adverse health effects of industrial wind turbines. Can Fam Physician (2013) 59:473–5.

52. Jeffery RD, Krogh CME, Horner B. Industrial wind turbines and adverse health effects. Can J Rural Med (2014) 19:21–6.

53. Botha P. Wind turbine noise and health-related quality of life of nearby residents: a cross-sectional study in New Zealand. Proceedings of the 4th International Meeting on Wind Turbine Noise. Rome: INCE Europe (2011). p. 1–8.

54. Michaud DS, Keith SE, McMurchy D. Noise annoyance in Canada. Noise Health (2005) 7:39–47. doi:10.4103/1463-1741.31634

55. McMurtry RY. Toward a case definition of adverse health effects in the environs of industrial wind turbines: facilitating a clinical diagnosis. Bull Sci Technol Soc (2011) 31:316–20. doi:10.1177/0270467611415075

56. MassDEP and MDPH. Wind Turbine Health Impact Study: Report on Independent Expert Panel. Department of Environmental Protection and Department of Public Health (2012). Available from: http://www.mass.gov/dep/energy/wind/turbine_impact_study.pdf

57. Yuan H, Long H, Liu J, Qu L, Chen J, Mou X. Effects of infrasound on hippocampus-dependent learning and memory in rats and some underlying mechanisms. Environ Toxicol Pharmacol (2009) 28:243–7. doi:10.1016/j.etap.2009.04.011

58. Page LA, Petrie KJ, Wessely S. Psychosocial responses to environmental incidents: a review and proposed typology. J Psychosom Res (2006) 60:413–22. doi:10.1016/j.jpsychores.2005.11.008

59. Schwartz SP, White PE, Hughes RG. Environmental threats, communities, and hysteria. J Public Health Policy (1985) 6:58–77. doi:10.2307/3342018

60. Health Canada. Electric and Magnetic Fields from Power Lines and Appliances (Catalogue # H13-7/70-2012E-PDF). Ottawa: Government of Canada (2012).

61. WHO. Electromagnetic Fields. (2012). Available from: http://www.who.int/peh-emf/en/

62. ICNIRP (International Commission on Non-Ionizing Radiation Protection). Guidelines for limiting exposure to time-varying electric and magnetic fields (1 Hz–100 kHz). Health Phys (2010) 99:818–36. doi:10.1097/HP.0b013e3181f06c86

63. IARC (International Agency for Research on Cancer). Working Group on the Evaluation of Carcinogenic Risks to Humans. Nonionizing Radiation, Part 1: Static and Extremely Low-Frequency (ELF) Electric and Magnetic Fields. (Monographs on the Evaluation of Carcinogenic Risks to Humans, 80). Lyon: IARC (2002).

64. National Institute of Environmental Health Sciences. EMF-Electric and Magnetic Fields Associated with the Use of Electric Power. Questions & Answers. Research Triangle Park, NC: NIEHS/DOE EMF RAPID Program (2002).

65. WHO. Electromagnetic Fields and Public Health, Exposure to Extremely Low Frequency Fields Fact Sheet No. 322. Geneva: World Health Organization (2007).

66. EPA (U.S. Environmental Protection Agency). EMF in Your Environment: Magnetic Field Measurements of Everyday Electrical Devices. Washington, DC: Office of

Radiation and Indoor Air, Radiation Studies Division, U.S. Environmental Protection Agency (1992).

67. UK DECC. Update of UK Shadow Flicker Evidence Base: Final Report. London: Department of Energy and Climate Change (2011).

68. Haugen KMB. International Review of Policies and Recommendations for Wind Turbine Setbacks from Residences: Setbacks, Noise, Shadow Flicker, and Other Concerns. St. Paul, MN: Minnesota Department of Commerce (2011). p. 1–43.

69. Intrinsik. Scientific Critique of Articles BSTS 2011, Vol 31. Final report (2011).

70. Phillips CV. Properly interpreting the epidemiologic evidence about the health effects of industrial wind turbines on nearby residents. Bull Sci Technol Soc (2011) 31:303–15. doi:10.1177/0270467611412554

71. Harrison JP. Wind turbine noise. Bull Sci Technol Soc (2011) 31:256–61. doi:10.1177/0270467611412549

72. Krogh CME. Industrial wind turbine development and loss of social justice? Bull Sci Technol Soc (2011) 31:321–33. doi:10.1177/0270467611412550

73. Thorne B. The problems with "noise numbers" for wind farm noise assessment. Bull Sci Technol Soc (2011) 31:262–90. doi:10.1177/0270467611412557

74. Bronzaft AL. The noise from wind turbines: potential adverse impacts on children's well-being. Bull Sci Technol Soc (2011) 31:291–5. doi:10.1177/0270467611412548

75. Shain M. Public health ethics, legitimacy, and the challenges of industrial wind turbines: the case of Ontario, Canada. Bull Sci Technol Soc (2011) 31:346–53. doi:10.1177/0270467611412552

76. Salt AN, Kaltenbach JA. Infrasound from wind turbines could affect humans. Bull Sci Technol Soc (2011) 31:296–302. doi:10.1177/0270467611412555

77. National Health and Medical Research Council in Australia. Wind Turbines and Health: A Rapid Review of the Evidence. Canberra, ACT: Commonwealth of Australia (2010). p. 1–11.

78. Chief Medical Officer of Health Ontario. The Potential Health Impact of Wind Turbines. Chief Medical Officer of Health (CMOH) report. Toronto, ON: Queen's Printer for Ontario (2010). p. 1–14.

79. Oregon Health Authority. Strategic Health Impact Assessment on Wind Energy Development in Oregon. Salem, OR: Office of Environmental Public Health, Public Health Division (2013).

80. Merlin T, Newton S, Ellery B, Milverton J, Farah C. Systematic Review of the Human Health Effects of Wind Farms. Canberra, ACT: National Health and Medical Research Council (2014).

81. Berglund B, Lindvall T editors. Community Noise. Stockholm: Center for Sensory Research, Stockholm University and Karolinska Institute (1995).

82. Laszlo HE, McRobie ES, Stansfeld SA, Hansell AL. Annoyance and other reaction measures to changes in noise exposure – a review. Sci Total Environ (2012) 435:551–62. doi:10.1016/j.scitotenv.2012.06.112

83. WHO. Burden of Disease from Environmental Noise: Quantification of Healthy Life Years Lost in Europe. Copenhagen: WHO Regional Office for Europe (2011).

84. Grimwood CJ, Skinner GJ, Raw GJ. The UK national noise attitude survey 1999/2000. Proceedings of the Noise Forum Conference; 2002 May 20; London: CIEH (2002).

85. Babisch W. The noise/stress concept, risk assessment and research needs. Noise Health (2002) 4:1–11.

86. Fields JM, de Jong RG, Gjestland T, Flindell IH, Job RFS, Kurra S, et al. Standardized general-purpose noise reaction questions for community noise surveys: research and recommendation. Sound Vib (2001) 242:641–79. doi:10.1006/jsvi.2000.3384

87. Fields JM, de Jong R, Brown AL, Flindell IH, Gjestland T, Job RFS, et al. Guidelines for reporting core information from community noise reaction surveys. Sound Vib (1997) 206:685–95. doi:10.1006/jsvi.1997.1144

88. Job RFS. The role of psychological factors in community reaction to noise. In: Vallet M editor. Noise as a Public Health Problem. (Vol. 3), Arcueil Cedex: INRETS (1993). p. 47–79.

89. Öhrström E. Longitudinal surveys on effects of changes in road traffic noise. J Acoust Soc Am (2004) 115:719–29. doi:10.1121/1.1639333

90. Öhrström E, Skånberg A, Svensson H, Gidlöf-Gunnarsson A. Effects of road traffic noise and the benefit of access to quietness. J Sound Vib (2006) 295:40–59. doi:10.1016/j.jsv.2005.11.034

91. Fields JM. Effect of personal and situational variables on noise annoyance in residential areas. J Acoust Soc Am (1993) 93:2753–63. doi:10.1121/1.405851

92. Guski R. Personal and social variables as co-determinants of noise annoyance. Noise Health (1999) 1:45–56.

93. Miedema HME, Vos H. Demographic and attitudinal factors that modify annoyance from transportation noise. J Acoust Soc Am (1999) 105:3336–44. doi:10.1121/1.424662

94. Boss LP. Epidemic hysteria: a review of the published literature. Epidemiol Rev (1997) 19:233–43. doi:10.1093/oxfordjournals.epirev.a017955

95. Henningsen P, Priebe S. New environmental illnesses: what are their characteristics? Psychother Psychosom (2003) 72:231–4. doi:10.1159/000071893

96. Petrie KJ, Sivertsen B, Hysing M, Broadbent E, Moss-Morris R, Eriksen HR, et al. Thoroughly modern worries: the relationship of worries about modernity to reported symptoms, health and medical care utilization. J Psychosom Res (2001) 51:395–401. doi:10.1016/S0022-3999(01)00219-7

97. Deignan B, Harvey E, Hoffman-Goetz L. Fright factors about wind turbines and health in Ontario newspapers before and after the Green Energy Act. Health Risk Soc (2013) 15:234–50. doi:10.1080/13698575.2013.776015

98. American Psychological Association (APA). Stress in America Findings. Washington, DC: APA (2010).

99. Mayo Clinic. Stress symptoms: Effects on Your Body, Feelings and Behavior. (2011). Available from: http://www.mayoclinic.com/health/stress-symptoms/SR00008_D

100. WHO. Night Noise Guidelines for Europe. Copenhagen: WHO Regional Office for Europe (2009).

101. McCoy D, Montgomery H, Arulkumaran S, Godlee F. Climate change and human survival. BMJ (2014) 348:g2351. doi:10.1136/bmj.g2351

# CHAPTER 11

# Public Engagement with Large-Scale Renewable Energy Technologies: Breaking the Cycle of NIMBYism

PATRICK DEVINE-WRIGHT

## 11.1 INTRODUCTION

Over the past few years, economists and climate scientists have constructed a strong evidence base to show that rising levels of atmospheric greenhouse gas are causing significant changes to global weather patterns. [1,2] In most developed countries, the energy required for transport, heat, and power is derived predominantly from greenhouse gas emitting fossil fuel sources (i.e. natural gas, coal, and oil). To mitigate climate change, governments are making commitments to reduce reliance upon these sources of energy and increase the use of low-carbon sources such as nuclear and renewable energy (e.g., solar, wind, and marine). During 2008, at least 73 countries set policy targets to increase the use of renewable energy, up from 66 in 2007. [3] For example, the UK government aims to increase the proportion of electricity generated from renewable energy sources from a current level of 5.5% to 30% by 2020, [4] as part of a 'step change'

*Reprinted with permission from the author and John Wiley & Sons. Devine-Wright P. Public Engagement with Large-Scale Renewable Energy Technologies: Breaking the Cycle of NIMBYism. Wiley Interdisciplinary Reviews: Climate Change 2,1 (2011). DOI: 10.1002/wcc.89.*

programme to achieve an 80% reduction in greenhouse gas emissions by 2050. [5]

Policy targets for substantial and rapid increases in the use of renewable energy are ambitious, particularly in the UK, given the controversies that have often surrounded large-scale renewable energy projects in the past, particularly onshore wind farms. These have involved bitter disputes between developers and affected communities, leading to projects being delayed and even abandoned [6,7] and opponents being dubbed 'NIMBYs' (not in my backyard). [8] For this reason, the subject of public engagement with renewable energy is an important one that may well play a crucial role in determining whether energy targets will be achieved both in the UK and elsewhere. Moreover, I have serious doubts that 'step changes' are practically achievable in such a short time frame, arising from the ways that policy makers and industry typically conceive the social aspects of system change, and particularly issues of public engagement.

Focusing upon renewable energy, this article aims to critically review recent policy statements and research findings to elucidate how engagement is conceived and practiced by policy makers and developers, with a specific focus upon the UK. I aim to briefly outline some of the problems with current approaches and suggest potential solutions. Analyses of public engagement with other aspects of climate mitigation and adaptation, including with nuclear power or geoengineering fall outside the scope of this article.


## 11.2 PUBLIC ENGAGEMENT

Over the past 20 years, public engagement has become a keystone of many different sectors of policy and decision making, encompassing the environment and sustainable development, science and technology, spatial planning, and more recently climate change. Internationally, examples include the Aarhus convention which enshrined a legal commitment upon states to enable public 'access to information, participation in decision-making, and access to justice on environmental matters'. [9] In the UK, the most recent Planning White Paper states that 'there must be full and fair opportunities for public consultation and community engagement'. [10]

Despite the growing conventionality of public engagement (PE) in environmental policy making and planning, it remains fraught with ambiguity, tensions, and dilemmas. PE is often used synonymously with the concept of public participation, making it difficult to distinguish between divergent forms of engagement. Rowe and Frewer [11] distinguished three foci: communication, consultation, and participation. These differ by the flow of information occurring between the parties and by its significance in the policy- or decision-making process. Communication involves one-way information flow from the 'sponsor' to the public where feedback is not sought; consultation may involve two-way information flow but the information flows back without any dialogue, while participation involves a two-way exchange of information between sponsor and public with the possibility for transformed opinions in both parties. Mechanisms of PE include leaflets, websites, phonelines and questionnaires (communication), public meetings and exhibitions (consultation), and citizens forums, public dialogues and deliberative polls (participation). [12]

Why engage with the public? The literature has identified three predominant rationales, succinctly defined by Stirling [13] as normative, substantive, and instrumental: 'Under a normative view, participation is just the right thing to do. From an instrumental perspective, it is a better way to achieve particular ends. In substantive terms, it leads to better ends' (Ref 13, p. 220).

Normative and substantive rationales for PE are central elements of sustainable development and the formation of sustainable communities, specifically claims that citizens in a democracy should be empowered to participate in decision-making processes, that resulting decisions are better quality, that citizens' consent and trust can be gained from PE in addition to fostering change to attitudes and behaviors. [14–16] Thus, a sustainability perspective would generally opt to open-up discussions by employing participatory, two-way mechanisms of public engagement, rather than close-down debate through the use of one-way mechanisms of communication. Recognition of these diverse rationales for PE is important, as the same engagement initiative may be instigated or supported by organizations holding quite different rationales which, left implicit, can create tensions and difficulties. [17]

Academics have criticized the beliefs held by actors instigating engagement initiatives about the public, which typically underlie attempts to restrict engagement to one-way communication informed by a purely instrumental rationale. [13] Researchers have identified a 'deficit model' that presumes lay people, in contrast to experts, to be ignorant of technical issues and unable or unwilling to engage with policies around new technologies. [18] Moreover, use of the concept 'the public' (as well as related terms such as 'individuals' or 'communities') can easily lead to an overtly homogeneous conception of the people targeted by specific initiatives, underplaying their diversity and overlooking marginalized groups. [19,20] Although such critiques have led some to advocate greater levels of public participation 'upstream' in science policy making and technological deployment, [21] there has also been criticism of 'extravagant' expectations of upstream engagement, [22] with such actions themselves often 'closing down' rather than 'opening up' debate [13] and undermining rather than building public trust. [23]

## 11.3 PUBLIC ENGAGEMENT, CLIMATE CHANGE, AND RENEWABLE ENERGY

Mitigating climate change involves both reducing energy consumption and increasing the use of renewable energy sources. Regarding consumption, the UK government has developed a novel framework for understanding environmental behavior, [24] identifying 14 behaviors targeted for change, different segments of 'the public', and a model of behavioral change that could help individuals 'to understand their role and responsibility in tackling climate change' (Ref 24, p6). Mechanisms of engagement employed by the government have chiefly consisted of information packages (including internet-based advice services), communication initiatives to raise awareness, and economic incentives to foster behavioral change. Allied to these, annual questionnaire surveys of nationally representative samples are commissioned to establish whether beliefs and self-reported behaviors change over time.

Renewable energy encompasses wind, biomass, hydro, solar, and marine energy. Unlike nuclear energy, these can be exploited at a variety of scales from small-scale building-level solar panels to large-scale hydro-electric schemes and offshore wind farms that produce sufficient electricity for hundreds of thousands of homes. In terms of public engagement with renewable energy, UK policy bears similarity to climate policies in declaring 'a role for everyone' (Ref 25, p. 151). However, a close scrutiny of how this role is conceived by policy makers at different scales of technology deployment reveals important distinctions. At smaller scales, households and communities are conceived to play an active role in increasing renewable energy use; as a result they are the target of measures to encourage wider adoption, including the provision of one-way communications (e.g., information and advice campaigns) and financial incentives (e.g., feed-in tariffs). At this small-scale, members of the public are conceived as consumers acting singly or in consort that will adopt voluntarily a range of new, unfamiliar, and rather expensive technologies to generate electricity or heat, providing they receive sufficient encouragement to enable them to do so.

At larger scales, renewable energy policy refers to the public in a different manner: as communities that 'host' renewable energy projects instigated by private developers. This passive role is played out in policy emphasis upon the activities of private developers: 'Renewable energy developers also have a central role to play in building local support for their projects, by ensuring effective engagement with local communities and sharing some of the benefits from renewable deployment with those who host them' (Ref 25, p. 154).

There is little to suggest that policy makers value pro-active engagement by the public (i.e., 'communities') in larger scale renewable energy projects, either from the normative rationale of seeing such projects as vehicles for fostering sustainable communities, or the substantive rationale of achieving better quality technology projects. Instead, the main focus of public engagement in larger-scale schemes is to secure public acceptance of developer-led projects.

This distinction between different scales is mirrored in recent engagement practices instigated by the UK government. Leading up to the publication of the Renewable Energy Strategy in 2009, the Department for Energy and Climate Change conducted a 'public dialogue' (the 'Big Energy Shift') with households in England, Wales, and Northern Ireland. This was a two-way engagement initiative that brought together small groups of householders, diverse stakeholders, and senior policy makers. When interviewed as part of the evaluation process, householders expressed appreciation for the opportunity to question technical experts about the merits of unfamiliar technologies, as well as a confidence that the dialogue would have an impact upon policy making, owing to the ways that senior politicians (including the Minister for Energy and the Secretary of State for Climate Change) participated in small-group meetings and listened to their views. [26]

However, the focus of the dialogue was solely upon actions at the household and community levels, overlooking larger-scale actions to reduce greenhouse gas emissions. The public dialogue has been followed by a practical initiative known as the Low Carbon Communities Challenge, involving £10 million of grant funding. Despite a more communitarian perspective upon behavioral change and technology deployment, the focus of the initiative continues to suggest that active public involvement in increasing renewable energy is only encouraged by policy makers at smaller scales of deployment.

## 11.4 THE INFLUENCE OF NIMBYISM

I believe the reason for this absence of encouragement is due to the widespread, yet implicit, influence of a particularly negative way of thinking about public engagement—that of NIMBYism. The NIMBY label ('not in my backyard') is a succinct way of pejoratively describing local opposition to unwanted land uses, attributing hostility to the ignorance, irrationality, or prejudice of members of the public. [8] NIMBYism blames technology controversies upon local objectors; homogenizes and simplifies their arguments [27]; and counterbalances an often poorly defined 'national interest' in developing large-scale infrastructure against 'local concerns'. [28]

Intrinsic to NIMBYism is the 'deficit' view of the public, [18] presuming that lay people are ignorant of technical issues and unable or unwilling to engage with policies around new technologies. NIMBYism is also a way of thinking about how people relate to the environment, presuming that residents' concerns are restricted to immediate, privately owned property and stem from self-interested values. [29] Thus, NIMBYism suggests a territorially bound view of local environments (the 'backyard'), despite the increasing interconnection between localities in a globalized world. [30] NIMBYism has been strongly critiqued by academic social scientists as an inaccurate and unfair way of describing local opposition. [31–33] However, it remains a popular way of thinking about local opposition, However, it remains a popular way of thinking about local opposition, which is often presumed in media commentaries to lead to delays in planning, [34] as has been shown by a recent study. [35]

Although not explicitly mentioned in UK renewable energy policy documents, there is both anecdotal and research evidence to suggest that NIMBYism plays an implicit role in shaping the ways that decision makers think about and undertake public engagement. For example, the Secretary of State for Energy and Climate Change was widely reported in the media in 2009 as attacking objections to wind farm proposals as 'socially unacceptable', recommending that society should regard objectors in a similar manner to car drivers who choose not to wear seat belts or stop at pedestrian crossings. Similarly, another senior government politician called for 'Age of Stupid' awards to be given out to those who object to wind farms. [36]

Recent changes to planning law also suggest a negative conception of public engagement with large-scale energy projects. The 2008 Planning Act aimed to streamline and speed up planning, removing public inquiries; and devised an Infrastructure and Planning Commission to make decisions on nationally significant infrastructure projects. This has resulted in a closing down of 'institutional spaces for challenging the status quo' (Ref 37, p. 405), spatially distancing political arenas for decision making from the localities directly affected by them, and restricting the ability of local residents to question not only the merits of specific proposals, but also the 'national interest' lying behind government policies. [28]

Recent research has shown that private sector developers hold similar views. As part of the 'Beyond Nimbyism' research project (for more info, see the Beyond Nimbyism project website), in-depth interviews revealed that developers consider members of the public to be an 'ever-present danger' who could at any moment act to obstruct their proposals, a situation that required the provision of appropriate information to allay public 'concerns' and 'fears'. [38] Allied to this, developers held a deep-seated antipathy toward the expression of emotion, with protestors viewed as emotional individuals capable of swaying what should be an exclusively 'rational' decision-making process in which facts should predominate. [39]

These conceptions of the public have several important consequences. First, they have helped to shape the development trajectories of specific technologies. For example, at least some of the motivation to promote wind energy offshore (despite increased costs and technical challenges in comparison to onshore) has arisen from the presumption that distancing the technology from 'backyards' will lessen 'NIMBY' opposition. [38,40] Second, conceptions of the public have informed which engagement mechanisms are used when new projects are proposed. In the eight case studies of renewable energy projects conducted in the 'Beyond Nimbyism' research project (ranging across bio-energy, marine, onshore, and offshore wind sectors), one-way communication predominated, involving websites, leaflets, phonelines, and questionnaire surveys. Although some two-way consultation was undertaken, it was carefully managed. Developers avoided the use of public meetings to prevent opponents having the opportunity to collectively express emotional antagonism and influence other citizens, and preferred public exhibitions as they afforded one-to-one engagement with individuals. [41] Such conceptions of the public have led to a growing focus upon providing 'community benefits packages' for each project, which has become a normative practice among wind energy developers over the past 5 years, and which aim to allay the presumed self-interested concerns of local residents. [42,43]

Walker et al. [44] draws attention to the interplay of different aspects of public engagement over time. Conceptions of the public held by decision makers lead to specific engagement mechanisms being used to engage with the public, and which in turn can lead to oppositional responses by the public, which feedback into the conceptions held by decision makers. In essence, there appears to be a rather destructive, self-fulfilling cycle at play in which local opposition is interpreted by developers and policy makers as evidence of NIMBYism, which leads to engagement practices whose main goal is to allay NIMBY responses (e.g., by limiting opportunities to participate, remedying information deficits and addressing self-interested concerns), but which in turn lead to more local opposition (e.g., arising from discontent at limited opportunities to participate, the invalidation of emotional response and the pre-occupation with financial benefits), which is then interpreted by developers as evidence of NIMBYism, and so on (Figure 1).

The value of NIMBY thinking is further undermined by the fact that research into the beliefs held by local residents about proposed projects fail to support key assumptions. For example, analysis conducted in the Beyond Nimbyism project on data from questionnaires distributed to residents in eight case studies (n = 2911) revealed no relation between key aspects of NIMBYism, such as perceived proximity to the project site or length of residence in the area, and residents' levels of support for project proposals. [45] Such findings tally with previous studies that have failed to provide empirical support for key tenets of NIMBYism. [46,47]

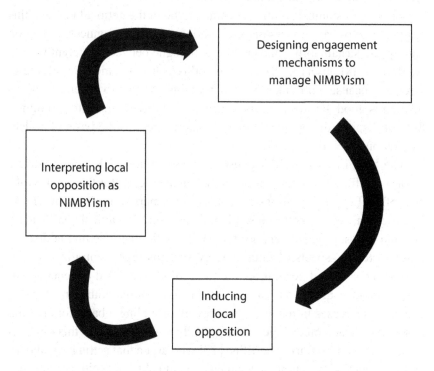

**FIGURE 1:** The cycle of NIMBYism in public engagement with renewable energy.

## 11.5 BREAKING OUT OF THE NIMBY CYCLE

The cycle of action and reaction in public engagement depicted above does not provide a suitable social context for the rapid transformation implied by the 30% target. A 'step change' increase in renewable energy is as much a transformation in thinking as it is a technical, economic, or institutional transformation. Habitual assumptions that may have proved useful in the past may become counter-productive in the future, no longer 'fit for purpose'. Here, I sketch an alternative conception that recasts 'how' and 'where' public engagement takes place, interlinking temporal processes and spatial outcomes.

To achieve a more sustainable energy system requires a process that promotes active public engagement at both smaller and larger scales of technology deployment. Rather than seeking acceptance by the public of pre-ordained technical solutions deemed to be in the national interest, this requires a two-way process of participation that better connects policy on energy and sustainability and enables a dialogue between different values. As Ratner [48] observed 'The sustainability concept is meaningful, therefore, not because it provides an encompassing solution to different notions of what is good, but for the way it brings such differences into a common field of dispute, dialogue, and potential agreement as the basis of collective action' (Ref 48, p. 62).

The 'Big Energy Shift' suggests openness among UK policy makers toward a different conception of public engagement. But the main weakness of the Big Energy Shift was that it was too narrow in focus, overlooking interconnections between small and large scale technical deployment, demand and supply side energy issues. A public dialogue that addressed more systemic aspects of future energy systems (e.g., centralized vs decentralized scales of generation, the 'mix' of sources used to generate energy, prices, or quotas for carbon) better matches the ambition of rapid, extensive increases in renewable energy by providing a basis for creating a 'social contract' on technical change. A dialogue process of this kind has been advocated by Barry and Ellis, [27] who argue that setting mandatory carbon reduction targets at the local level would oblige the citizens of every municipality to engage with energy/climate change while simultaneously offering flexibility over the most appropriate ways such a target could be met. But such a process cannot be solely local, as large-scale projects have outcomes that transcend localities [28]; thus requiring a more effective integration between national and local levels than currently exists in order to ensure coherence between multiple 'bottom up' activities.

Such a process not only challenges existing ways of thinking about how public engagement occurs, it challenges assumptions about where engagement should take place. Analysis has indicated that policy makers and developers view the contexts of large-scale projects either as 'sites' to be developed or 'backyards' to be avoided. [29] Conceiving localities as sites strips places of the layers of subjective meanings and emotions that are an important element of their character. [49,50] Seeing localities as backyards reinforces NIMBY presumptions that public responses stem solely from self-interested concerns with private property. [29] Furthermore, changes to 'sites' or 'backyards' are commonly conceived in zero-sum terms, pitting the 'global' benefits of renewable energy against 'local' costs. [40] This way of thinking is also prevalent in renewable energy policy, [25] as the emphasis upon benefit sharing is founded upon a similar cost-benefit rationale. At an extreme, this perspective views the destruction of local places as a necessary step to avoid global climatic change—what has been dubbed a discourse of 'place sacrifice'. [7]

Conceiving spatial aspects of public engagement in these ways is problematic because it overlooks important aspects of how people and places connect. Localities are not just 'sites' that can be objectively assessed and altered by experts, but are 'places' that residents feel emotionally attached to, and which can become an important element of their sense of identity. [51] Empirical research has shown how proposals for large-scale renewable energy projects—spatially distant offshore wind farms—can lead to public opposition because they are interpreted as a threat to place-related identities, leading to oppositional behavior. [52] This indicates the limitations of efforts by policy makers and developers to frame local environmental change in purely financial terms, and can lead to accusations of bribery. [43] What is required is a process of deployment that goes with the grain of place attachments and identities rather than erasing place meanings; one where aspects of place character that are important in sustaining local distinctiveness and historical continuity are identified from the outset of a development process, and ascribed similar importance in planning to more objective technical and environmental considerations. A more inclusive process that involves local residents in the emplacement of technologies in specific localities further supports the rationale for a dialogue process that can link places at different scales (local to regional to national to international), making the global local, but avoiding 'sacrificing' the local for the sake of the global, aiming through public dialogue to transcend zero-sum outcomes.

## 11.6 CONCLUSION

This review revealed how UK policy makers promote active public engagement with renewable energy only at smaller scales of technology deployment, preferring a more passive role at larger scales, influenced by NIMBYism. NIMBY conceptions presume the public to be an 'ever present danger', arising from deficits in factual knowledge or surfeits of emotion, to be marginalized through streamlined planning processes and one-way engagement. Despite this negative conception, there is widespread social consent for increasing the use of renewable energy, including wind energy, both in the UK and internationally, but it is fragile and qualified. Local opposition to large-scale projects is not inconsistent with this consent, but is a logical outcome of flawed conceptions and practices of public engagement. This is why breaking the cycle of NIMBYism is not merely an issue for the UK but for many countries (e.g., Netherlands, USA, Australia, and New Zealand) where local residents have contested development proposals [53–56]. To do so requires new ways of thinking about and practicing public engagement with energy system change, opening up a two-way public dialogue that better connects national policy making with the local places and residents most directly affected by large-scale schemes. Such a step would match the radical ambitions of rapid increases in renewable energy use with a process of change more likely to facilitate its achievement.

## REFERENCES

1.   Stern N. The Economics of Climate Change: The Stern Review. Cambridge, UK: Cambridge University Press; 2006.

2. Intergovernmental Panel on Climate Change (IPCC). Climate Change 2007: Synthesis Report. Geneva: IPCC; 2007.

3. REN21. Renewables Global Status Report: 2009 Update. Paris: REN21 Secretariat; 2009.

4. Department for Energy and Climate Change. The Low Carbon Transition Plan; 2009. Available at: http://www.decc.gov.uk/en/content/cms/publications/lc_trans_plan/lc_trans_plan.aspx. (Accessed May 3, 2010).

5. Committee on Climate Change (CCC). Meeting the Carbon Budgets: the Need for a Step Change. London: CCC; 2009.

6. Toke D. Explaining wind power planning outcomes, some findings from a study in England and Wales. Energy Policy 2005, 33:1527–1539.

7. Ellis G, Barry J, Robinson C. Many ways to say 'no', different ways to say 'yes': applying Q-methodology to understand public acceptance of wind farm proposals. J Environ Plann Manag 2007, 50:517–551.

8. Burningham K, Barnett J, Thrush D. The limitations of the NIMBY concept for understanding public engagement with renewable energy technologies: a literature review. School of Environment and Development, University of Manchester; 2006. Available at: http://geography.exeter.ac.uk/beyond_nimbyism/deliverables/outputs. shtml (Accessed May 3, 2010).

9. United Nations. Convention on Access to Information, Public Participation in Decision-making and Access to Justice in Environmental Matters, United Nations ECE/CEP/43, in Aarhus, Denmark, June 23–25, 1998.

10. HM Government. White Paper: Planning for a Sustainable Future. London: The Stationery Office; 2007.

11. Rowe G, Frewer LJ. A typology of public engagement mechanisms. Sci Technol Human Values 2005, 30:251–290.

12. Horlick-Jones T, Walls J, Rowe G, Pidgeon N, Poortinga W, Murdock G, O'Riordan T. The GM Debate: Risk, Politics and Public Engagement. Abingdon: Routledge; 2007.

13. Stirling A. Opening up or closing down? Analysis, participation and power in the social appraisal of technology. In: Leach M, Scoones I, Wynne B, eds. Science and Citizen: Globalization and the Challenge of Engagement. London: Zed; 2005, 218–231.

14. Selman P, Parker J. Citizenship, civicness and social capital in local agenda 21. Local Environ 1997, 2: 171–184.

15. Department for Environment, Food and Rural Affairs (DEFRA). Securing the Future—The UK Government Sustainable Development Strategy. London: DEFRA; 2005.

16. Department of Communities and Local Government. What is a Sustainable Community? Department of Communities and Local Government. London; 2007. Available at: http://www.communities.gov.uk/archived/general-content/communities/whatis/. (Accessed May 3, 2010).

17. Hoppner C. Public engagement in climate change—disjunctions, tensions and blind spots in the UK. IOP Conf Series Earth Environ Sci 2009, 8:012010. doi:10.1088/1755-1315/8/1/012010.

18. Wynne B. Rationality and Ritual: The Windscale Inquiry and Nuclear Decisions in Britain. Chalfont, St Giles: British Society for the History of Science; 1982.

19. Walker GP. Renewable energy and the public. Land Use Policy 1995, 12:49–59.
20. Renn O. Risk communication—consumers between information and irritation. J Risk Res 2006, 9:833–849.
21. Wilsdon J, Willis R. See-through Science: Why Public Engagement Needs to Move Upstream. London: Demos; 2004.
22. Wynne B. Risk as globalizing discourse? Framing subjects and citizens. In: Leach M, Scoones I, Wynne B, eds. Science and Citizens: Globalization and the Challenge of Engagement. London/New York: Zed Books; 2005.
23. Petts J. Public engagement to build trust: false hopes? J Risk Res 2008, 11:821–835.
24. Department for Environment, Food and Rural Affairs (DEFRA). A Framework for Pro-Environmental Behaviours. London: DEFRA; 2008.
25. Department for Energy and Climate Change. The UK Renewable Energy Strategy; 2009. Available at: http://www.decc.gov.uk/en/content/cms/what_we_do/uk_supply/energy_mix/renewable/res/res.aspx. (Accessed May 3, 2010).
26. Rathouse K, Devine-Wright P. Evaluation of the Big Energy Shift. Report to DECC and Sciencewise; 2010.
27. Barry J, Ellis G. Beyond consensus? Agonism, republicanism and a low carbon future. In: Devine-Wright P, ed. Public Engagement with Renewable Energy: From NIMBY to Participation. London: Earthscan; 2010, 29–42.
28. Owens S. 'A collision of adverse opinions'? Major projects, planning inquiries, and policy change. Environ Plann A 2002, 34:949–957.
29. Devine-Wright P. From backyards to places: public engagement and the emplacement of renewable energy technologies. In: Devine-Wright P, ed. Public Engagement with Renewable Energy: From NIMBY to Participation. London: Earthscan; 2010, 57–70.
30. Castells M. Informationalism and the network society. In: HimanenP, ed. The Hacker Ethic and the Spirit of the Information Age. New York: Random House; 2001, 155–178.
31. Burningham K. Using the language of NIMBY: a topic for research not an activity for researchers. Local Environ 2000, 5:55–67.
32. Wolsink M. Invalid theory impedes our understanding: a critique on the persistence of the language of NIMBY. Trans Inst Br Geogr 2005, 31:85–91.
33. McClymont K, O'Hare P. "We're not NIMBYs!" contrasting local protest groups with idealised conceptions of sustainable communities. Local Environ 2008, 13:321–335.
34. Toynbee P. NIMBYs can't be allowed to put a block on wind farms. The Guardian, January 5, 2007.
35. Aitken M, McDonald S, Strachan P. Locating 'power' in wind power planning processes: the (not so) influential role of local objectors. J Environ Plann Manag 2008, 51:777–799.
36. Charlesworth A. Prescott proposes stupidity awards for wind farm 'nimbys' Business-Green; 2009. Available at: www.businessgreen.com/business-green/news/2248512/prescott-proposes-stupidity. (Accessed May 3, 2010).
37. Cowell R, Owens S. Governing space: planning reform and the politics of sustainability. Environ Plann C Gov Policy 2006, 24:403–421.

38. Walker G, Cass N, Burningham K, Barnett J. Renewable energy and sociotechnical change: imagined subjectivities of 'the public' and their implications. Environ Plann A 2010, 42:931–947.

39. Cass N, Walker G. Emotion and rationality: the characterisation and evaluation of opposition to renewable energy projects. Emotion Space Soc 2010, 2:62–69.

40. Haggett C. Over the sea and far away? A consideration of the planning, politics and public perception of offshore wind farms. J Environ Policy Plann 2008, 10:289–306.

41. Barnett J, Burningham K, Walker G, Cass N. Imagined publics and engagement around renewable energy technologies in the UK. Public Underst Sci. doi: 10.1177/0963662510365663.

42. Renewables Advisory Board. Delivering community benefits from wind energy development: a toolkit; 2009. Available at: http://www.renewables-advisory-board. org.uk/vBulletin/showthread.php?p=221. (Accessed May 3, 2010).

43. Cass N, Walker G, Devine-Wright P. Good neighbours, public relations and bribes: the politics and perceptions of community benefit provision in renewable energy development. J Environ Policy Plann. 2010, 12:255–275.

44. Walker G, Devine-Wright P, Barnett J, Burningham K, Cass N, Devine-Wright H, Speller G, Barton J, Evans B, Heath Y, et al. Symmetries, expectations, dynamics and contexts: a framework for understanding public engagement with renewable energy projects. In: Devine-Wright P, ed. Public Engagement with Renewable Energy: From NIMBY to Participation. London: Earthscan. 2010, 1–14.

45. Devine-Wright P, Walker G, Barnett J, Burningham K, Devine-Wright H, Evans B, Infield D, Speller G, Cass N, Barton J, et al. Overall Project Summary Report; 2009. Available at: http://geography.exeter.ac.uk/beyond_nimbyism/deliverables/reports. shtml (Accessed May 3, 2010).

46. Jones C, Eiser JR. Understanding 'local' opposition to wind development in the UK: how big is a backyard? Energy Policy 2010, 37:4604–4614.

47. Michaud K, Carlisle JE, Smith E. NIMBYism vs. environmentalism in attitudes towards energy development. Env Polit 2008, 17:20–39.

48. Ratner BD. Sustainability as a dialogue of values: challenges to the sociology of development. Sociol Inq 2004, 74:50–69.

49. Tuan YF. Space and Place: The Perspective of Experience. Minneapolis: University of Minnesota Press; 1977.

50. Easthorpe H. A place called home. Housing Theory Soc 2004, 21:128–138.

51. Devine-Wright P. Rethinking Nimbyism: the role of place attachment and place identity in explaining place protective action. J Community Appl Soc Psychol 2009, 19:426–441.

52. Devine-Wright P, Howes Y. Disruption to place attachment and the protection of restorative environments: a wind energy case study. J Environ Psychol. doi:10.1016/j. jenvp.2010.01.008. Forthcoming.

53. Wolsink M. Wind power implementation: the nature of public attitudes: equity and fairness instead of 'backyard motives'. Renew Sustain Energ Rev 2007, 11:1188–1207.

54. Swofford T, Slattery M. Public attitudes to wind energy in Texas: local communities in close proximity to wind farms and their effect on decision-making. Energy Policy 2010, 38:2508–2519.

55. Gross C. Community perspectives of wind energy in Australia: the application of a justice and community fairness framework to increase social acceptance. Energy Policy 2007, 35:2727–2736.
56. Graham JB, Stephenson JR, Smith IJ. Public perceptions of wind energy developments: case studies from New Zealand. Energy Policy 2009, 37:3348–3357.

# PART III

# TRENDS AND FUTURE CHALLENGES

# CHAPTER 12

# Wind Turbine Condition Monitoring: State-of-the-Art Review, New Trends, and Future Challenges

PIERRE TCHAKOUA, RENÉ WAMKEUE, MOHAND OUHROUCHE, FOUAD SLAOUI-HASNAOUI, TOMMY ANDY TAMEGHE, AND GABRIEL EKEMB

## 12.1 INTRODUCTION

Energy conversion and efficiency improvement have become a worldwide priority to secure an energy supply and address the challenges of climate change, greenhouse gas emission reduction, biodiversity protection, and renewable technology development. In 2011, renewable sources accounted for nearly 50% of the estimated globally added electric capacity evaluated at 208 GW [1]. Among all renewable energy sources, wind energy is the fastest-growing sector in terms of installed capacity. As shown in Figure 1, the cumulative installed wind power capacity reached 283 GW in 2011, which represents nearly 3% of global electricity production. Furthermore, the contribution of wind power to the world total generation capacity is expected to reach 8% by 2018 [1–3].

*Wind Turbine Condition Monitoring: State-of-the-Art Review, New Trends, and Future Challenges*
© *Tchakoua P, Wamkeue R, Ouhrouche M, Slaoui-Hasnaoui F, Tameghe TA, and Ekemb G. Current Status and Prospects of Biodiesel Production from Microalgae.* Energies *7,4 (2014), doi:10.3390/ en7042595. Licensed under Creative Commons Attribution 3.0 Unported License, http://creativecommons.org/licenses/by/3.0/.*

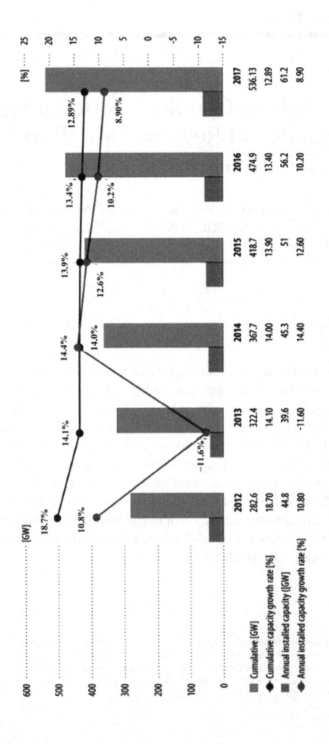

**FIGURE 1:** Wind energy world market forecast for 2013–2017 [1]. Reprinted/Reproduced with permission from [1]. Copyright 2013, Global Wind Energy Council (GWEC).

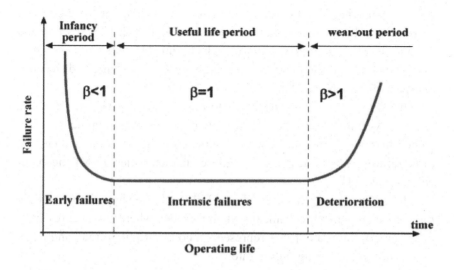

**FIGURE 2:** The "bathtub" curve illustrating the reliability of technical systems.

Wind turbines (WTs) are unmanned, remote power plants. Unlike conventional power stations, WTs are exposed to highly variable and harsh weather conditions, including calm to severe winds, tropical heat, lightning, arctic cold, hail, and snow. Due to these external variations, WTs undergo constantly changing loads, which result in highly variable operational conditions that lead to intense mechanical stress [4]. Consequently, the operational unavailability of WTs reaches 3% of the lifetime of a WT. Moreover, operation and maintenance (OM) costs can account for 10%–20% of the total cost of energy (COE) for a wind project, and this percentage can reach 35% for a WT at the end of life. A preventive-centered maintenance strategy that avoids machine shutdown can considerably reduce these costs [5–7]. Therefore, WTs require a high degree of maintenance to provide a safe, cost-effective, and reliable power output with acceptable equipment life. The state-of-the-art method for determining the maintenance strategy in the WT industry is reliability-centered maintenance (RCM), which consists of preventive maintenance based on performance and/or parameter monitoring and subsequent actions. In this strategy, condition-monitoring (CM) is used to determine the optimum point between corrective and scheduled maintenance strategies [8–11]. The recurrent and commonly used condition-monitoring techniques (CMTs) are: (i) vibration/acoustic-controlled and OM techniques for the turbine; and (ii) optical strain gauges for the blades.

The WTs are typically designed to operate for a period of 20 years [12,13]. As with other mechanical systems, time-based maintenance assumes that the failure behavior of WTs is predictable. Fundamentally, three failure patterns describe the failure characteristics of WT mechanical systems [14].

The bathtub curve shown in Figure 2 illustrates the hypothetical failure rate versus time in a mechanical system [15–18], where $\beta < 1$ represents a decreasing failure rate, $\beta = 1$ represents a constant failure rate, and $\beta > 1$ represents an increasing failure rate.

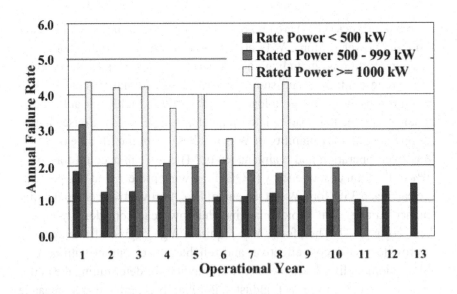

**FIGURE 3:** Number of incidents per wind turbine (WT) per operational year; WTs are categorized by rated power [12]. Reprinted/Reproduced with permission from [12]. Copyright 2008, American Society of Mechanical Engineers.

Guo et al. [19] developed a three-parameter Weibull failure rate function for WTs, and their results corroborate the bathtub curve. Echavarria et al. [12] published results of a remarkable 15-year research study on the frequency of failures versus increasing operational age for various WT power ratings (Figure 3).

The frequency of failures in WTs also varies with the scale and type. Spinato et al. [18,20] carried out a failure analysis based on onshore WT types, as specified in the Schleswig Holstein Landwirtschaftskammer (LWK) database. The work displayed a general trend of an increasing failure rate with turbine size. Because turbine capacity continues to grow, we can assume that it will be difficult to decrease the initial failure rate. Several research studies considered the distribution of WT failures in the main components [13,20,21]. Haln et al. [13] reported a survey of 1500 WTs over 15 years and found that five component groups, i.e., electrical system, control system, hydraulic system, sensors, and rotor blades, are responsible for 67% of failures in WTs, as shown by the pie chart in Figure 4.

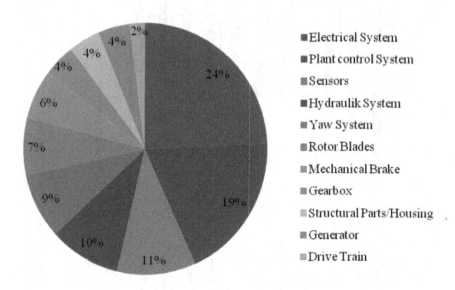

**FIGURE 4:** Share of the main components of the total number of failures [13]. Reprinted/ Reproduced with permission from [13]. Copyright 2007, Springer Science + Business Media.

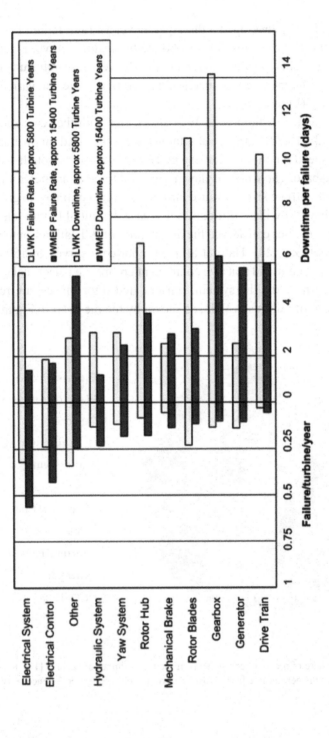

**FIGURE 5:** Failure rates and downtime from two large surveys of European WTs over 13 years [13]. Reprinted/Reproduced with permission from [13]. Copyright 2007, Springer Science + Business Media.

To establish the impact of component failure on WT reliability, research centered on the availability of WTs was presented in [8,22–25]. The results published by Fischer et al. [8] indicated that 75% of the annual downtime is caused by only 15% of the failures in WTs. This result corroborates the conclusions of Haln et al. [13], regarding the average failure rate and average downtime per component. The results of this study are also in agreement with the conclusions of Crabtree et al. [24], regarding the comparison of failure rates and downtime for different WT subassemblies based on surveys of European wind-energy conversion systems (WECSs). The chart in Figure 5 summarizes the failure rate and downtime of different WT subassemblies. The reliability and downtime data of the Egmond aan Zee wind farm in Germany also produced similar results, i.e., the gearbox failure rate is low but the downtime and resultant costs are high. As a result, the percentage of electricity production lost due to gearbox downtime is the highest of all subassemblies [24].

A statistical analysis of WT faults demonstrates that their reliability and availability depend on multiple factors, i.e., age, size, weather, wind speed, and subassembly failure rates. However, applying efficient CMTs can greatly increase the reliability of WTs.

In the literatures, few articles have provided a review of wind turbine condition monitoring (WTCM) and/or fault diagnosis [7,21,26–29]. The goal of this paper is to provide a review of methods and techniques for WTCM with a classification of: (i) intrusive and nonintrusive techniques; and (ii) destructive techniques and non-destructive techniques. This work also focuses on trends and future challenges in the WTCM industry. The paper is organized as follows: Section 2 is dedicated to CM-related concepts and definitions and outlines the relationships among CM, fault diagnosis, and fault prognostic and maintenance strategies; Section 3 presents a review of techniques and methods used in WECSs and CM, subdividing them into subsystem techniques and overall system techniques as well as destructive and non-destructive techniques; Section 4 discusses the new trends and future challenges that will enable the industry to address the WT challenges of the future, including reducing operational costs and improving reliability; finally, Section 5 provides conclusions to the work.

## 12.2 CONCEPTS AND DEFINITIONS

### 12.2.1 MAINTENANCE APPROACHES

As in most industries, maintenance approaches in the WT industry can be widely classified into three main groups [30,31]:

- Reactive or corrective maintenance (run to failure);
- Preventive maintenance (time-based);
- Predictive maintenance (condition-based).

The COE estimation for a WECS is given by Equation (1) [6,32–35], where ICC is the initial capital investment cost; FRC is the annual fixed charge rate; E is the annual energy production in kW h; and OM is the annual OM cost:

$$COE = \frac{ICC \cdot FCR + OM}{E} \tag{1}$$

where ICC and FRC are fixed parameters; and OM is a variable parameter that can affect the COE during the lifetime of the project. Therefore, the profit from wind energy is highly dependent on the ability to control and reduce this variable cost. The OM cost of equipment will notably depend on the maintenance strategy adopted by the user.

The cost associated with traditional maintenance strategies is presented in Figure 6 [30]. In a preventive maintenance strategy, the prevention cost will be quite high, whereas the repair cost will be low because many potential failures will not occur. In other words, preventive maintenance will considerably reduce the number of failures that occur but will be expensive. In a reactive maintenance strategy, a greater number of faults will occur and will lead to a high cost of repair and low cost of prevention. As shown on the graph, a combination of preventive and reactive maintenance strategies can improve the reliability, availability, and maintainability of WTs while simultaneously reducing the maintenance cost [4,6,30,36].

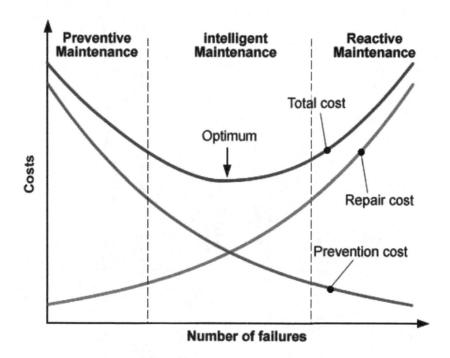

**FIGURE 6:** Costs associated with traditional maintenance strategies.

## 12.2 CM, DIAGNOSIS, AND MAINTENANCE THEORIES

Reliability is the ability of a device to perform the required functions under the given conditions for a given time [4,37]. The reliability of a WT is critical for extraction of the maximum energy available from wind. Reliability can be highly improved by the implementation of adequate condition-monitoring systems (CMSs) and fault detection systems (FDSs), and availability is a fundamental measure of reliability. Holen et al. [38] defined availability as the probability that a component or system is capable of functioning at time t, given by Equation (2), where MTTF is the mean time to failure and MTTR is the mean time to recovery:

$$A = \frac{MTTF}{MTTF + MTTR} \tag{2}$$

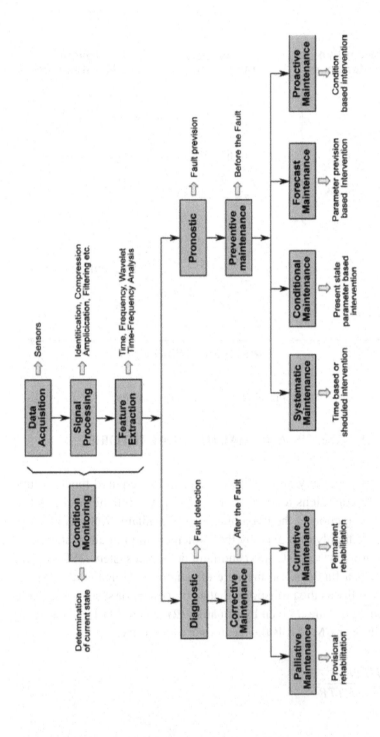

**FIGURE 7:** Overview of condition-monitoring (CM) and maintenance processes for WTs.

A CMS is a tool used to ensure and measure the reliability of any running system [39]. Wiggelinkhuizen et al. [40] suggested that for WECSs, significant changes are indicative of a developing failure. The continuous component states (i.e., WT health) are evaluated using a collection of techniques, i.e., vibration analysis (VA), acoustics, oil analysis (OA), strain measurement (SM), and thermography [27]. Data are sampled at regular time intervals using sensors and measurement systems. Using data processing and analyses, CMSs can determine the states of the key WECS components. By processing the data history, faults can be detected (diagnosis) or predicted (prognostic) and the appropriate maintenance strategy can be chosen.

Maintenance includes any actions appropriate for retaining equipment in or restoring it to a given condition [31]. Maintenance is required to ensure that the components continue to perform the functions for which they were designed. The basic objectives of the maintenance activity are to: (i) deploy the minimum resources required; (ii) ensure system reliability; and (iii) recover from breakdowns [41]. The applied maintenance strategy can be preventive if a predicted failure is avoided or corrective when a detected failure is repaired [42].

A description of and models for CMSs can be found in [27,39,43,44]. This description can be combined with concepts definitions provided in [14,31,45–48], which address maintenance techniques and methods. The diagram relating technical concepts and words used in the domain of WTCM and fault diagnosis emerges from the aforementioned combination. As shown in Figure 7, CM is performed in three main steps: data acquisition using sensors, signal processing using various data processing techniques, and feature extraction via the retrieval of parameters that will aid in establishing the current status of the monitored equipment. Using both: (i) current information sources; and (ii) information on the system's past status obtained from stored data, the system's present state is obtained via online monitoring such that a fault can be detected or predicted. After a fault is diagnosed, corrective maintenance is carried out. Two approaches to corrective maintenance can be distinguished, i.e., palliative maintenance, which consists of provisional solutions to failures, and curative maintenance for standing solutions to failures. If a fault is predicted, preventive maintenance is carried out before the fault can occur. In this case, four different approaches can be used: time-based or scheduled maintenance, current-state based or conditional maintenance, parameter-projection-based or forecasting maintenance, and status-based or proactive maintenance.

## 12.3 REVIEW OF CONCEPTS AND METHODS FOR WTCM

According to the Swedish standard SS-EN 13306 [49], monitoring can be defined as an activity performed either manually or automatically that is intended to observe the actual state of an item. The key function of a successful CMS should be to provide a reliable indication of the presence of a fault within the WECS and to indicate the location and severity of the situation [25]. For this purpose, a CMS is required for early warning sign detection. CM is based on data acquisition and signals processing and can be implemented using various approaches with different levels of technology [46].

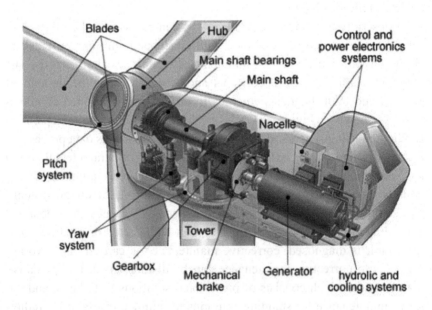

**FIGURE 8:** Typical main components of a utility-scale WT.

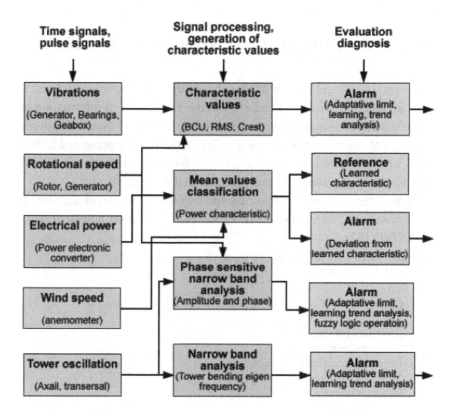

**FIGURE 9:** Function model for monitoring of a wind-energy conversion system (WECS) [53]. BCU: boundary controlling unstable; and RMS: root mean square. Reprinted/ Reproduced with permission from [53]. Copyright 2008, Blekinge Institute of Technology.

A complete CMS is composed of many subsystems, each monitoring a particular component of the wind generator [50]. Due to the considerable level of overlap between functions of different subsystems, certain CM subsystems will monitor many components of the WT. The approach proposed in this review differentiates CMTs applied on WT subsystems from CMTs applied on the overall WT system.

### 12.3.1 WT SUBSYSTEMS OR INTRUSIVE CM TECHNIQUES

The subsystem-level CM of WTs is based on subcomponents related to local parameters [27,28,51] and enables the acquisition of information on specific components and thus the precise localization of eventual failures. The typical main components of a utility-scale WT are presented in Figure 8, and an example of a function model for the monitoring of a WECS based on the subsystem approach is presented in Figure 9.

Subsystem CM can be classified into two main subcategories, namely, those based on destructive test (DT) and those based on non-destructive test (NDT) [52].

Subsystem CM based on DT uses:

- VA;
- OA;
- SM;
- Electrical effects;
- Shock pulse method (SPM);
- Physical condition of materials;
- Self-diagnosis sensors;
- Other techniques.

Subsystem CM based on NDT uses:

- Ultrasonic testing techniques (UTTs);
- Visual inspection (VI);
- Acoustic emission;
- Thermography;
- Performance monitoring;
- Radiographic inspection.

## 12.3.1.1 SUBSYSTEM CM TECHNIQUES BASED ON DTS

As stated in [54], a DT is "a form of mechanical test (primarily destructive) of materials whereby certain specific characteristics of the material can be evaluated quantitatively". DTs are generally realized more easily and yield additional data that are easier to interpret than those from NDTs [55,56]. As applied to WECSs, DTs are dynamic or static and can provide useful information related to the material's design considerations, equipment performance, structural health, and useful life.

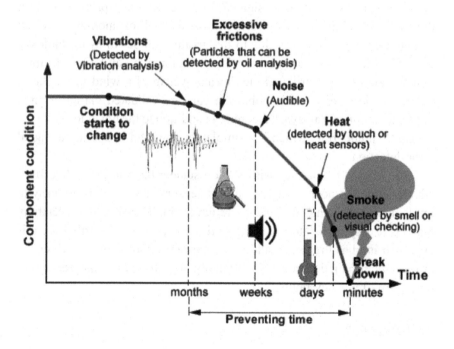

**FIGURE 10:** Typical development of a mechanical failure.

*12.3.1.1.1 VA*

VA is the most well-known technology for rotating equipment CM. As shown in Figure 10, VA is the most efficient technology for early prediction and detection of failures in mechanical equipment [32]. Applied sensor technology is selected by considering the frequency range and operating conditions [57]. Position transducers, velocity sensors, accelerometers, and spectral emission energy sensors are used for low-, middle-, high-, and very-high-frequency ranges, respectively. Fast Fourier transformation is the signal processing technique commonly used in VA to convert a time-domain signal into a frequency-domain signal [58].

As a subsystem monitoring technique, VA is applied to such WT components as shafts, bearings, gearboxes, and blades. In WTs or wind farms, CM via the VA's extreme false alarm levels can provide information on the incorrectness of vibration signals from the recording process (e.g., in the case of a faulty sensor). To minimize the risk of anomalies, which is increased in a wind farm due to the greater number of WTs, Jablosky et al. [59] developed an algorithm for the automatic validation of vibration signals in the distributed monitoring system of a wind farm. Based on amplitude validation, the vibration data are validated via an original implementation of Parseval's theorem, in addition to the novel idea of a so-called "N-point" rule, which is a simple yet powerful in automatic signal error detection.

The WTCM techniques with VA are standardized in ISO10816 [60], which define the positioning and use of sensors. VA methods are easy to implement in existing equipment and have a high level of interpretation, making it easy to locate the exact faulty component. Nevertheless, this approach implies the use of additional hardware and software, which increases the production costs. Additionally, it is difficult to use sensors to detect low-frequency faults [28].

*12.3.1.1.2 OA*

Oil debris monitoring has been proven as a viable CMT for the early detection and tracking of damage in bearing and gear elements in WT gearboxes [61]. Indeed, 80% of gearbox problems can be attributed to the bearings, which subsequently lead to damage to the gearing [62].

In most cases, oil is pumped through the component in a closed-loop system, and metal debris from cracked gearbox wheels or bearings is caught by a filter. The amount and type of metal debris can indicate the health of the component. OA has three main purposes [61]: (i) to monitor the lubricant; condition and reveal whether the system fluid is healthy and fit for further service or requires a change; (ii) to ensure the oil quality (e.g., contamination by parts, moisture); and (iii) to safeguard the components involved (part characterization). Six main tests are generally employed in the OA process: [57,63–65]:

- Viscosity analysis;
- Oxidation analysis;
- Water content or acid content analysis;
- Particle count analysis;
- Machine wear analysis;
- Temperature.

OA techniques can be divided into two categories: real-time continuous monitoring and offline oil sample analysis [66]. These processes are typically executed off line by taking samples. However, online real-time oil debris monitoring may be desirable for applications in which failure modes develop rapidly or when accessibility is limited. In this case, it is advisable to install several sensors in the gearbox lubrication loop to analyze different characteristics. This approach will increase the reliability and accuracy of the analyses [53,64].

The technology for on-line detection can be broadly divided into three subcategories depending on the sensing techniques applied [4]: electromagnetic sensing, flow or pressure-drop sensing, and optical debris sensing. In terms of cost, size, accuracy, and development, suitable oil monitoring technologies are online ferrography, selective fluorescence spectroscopy, scattering measurements, Fourier transform infrared (IR) spectroscopy, photo acoustic spectroscopy, and solid-state viscometry [62,64]. Du and Zhe [67] developed a high-throughout, high-sensitivity inductive sensor for the detection of micro-scale metallic debris in nonconductive lubrication oil. The device is able to detect and differentiate ferrous and non-ferrous metallic debris in lubrication oil with high efficiency.

Although OA is the only method for detecting cracks in the internal gearbox, this approach has two main limitations. First, it cannot detect failures outside the gearbox, and second, use of this equipment for online monitoring is highly expensive. For these reasons, offline monitoring of oil samples is often used [28,68].

### 12.3.1.1.3 Temperature Measurement (TM)

Monitoring the temperature of the observed component is one of the most common methods of CM [56]. TM aids in detecting the presence of any potential failure related to temperature changes in the equipment. In the wind energy industry, TM is applied on such components as bearings, fluids (oil), and generator windings, among others [53,69]. Optical pyrometers, resistant thermometers, and thermocouples are a subset of the sensors used in TM [70]. Unlike thermography, TM provides information on the ongoing deterioration process in the component from excessive mechanical friction due to faulty bearings and gears, insufficient lubricant properties, and loose or bad electrical connections [53].

TM is reliable because every piece of equipment has a limited operational temperature. However, temperature develops slowly and is not sufficient for early and precise fault detection [71]. Additionally, the measured temperature can also be influenced by the surroundings. Therefore, TM is rarely used alone but often as a secondary source of information. In this case, the primary source could be vibration monitoring [32,71].

### 12.3.1.1.4 SM

SM is a renowned technique for structural health monitoring (SHM) and is becoming increasingly important in the WT industry, where it is applied to blades and towers; SM is commonly used in laboratory settings for blade lifetime testing [16,51,72,73]. Measurements are gathered with sensors, i.e., so-called metal foil strain gauges, and the finite element method is commonly used to process the acquired data [73,74]. Strain gauges can be placed randomly on the blade, and the distribution varies according to the number of transducers. However, strain gauges are not robust over the long term, and more robust sensors might offer an interesting application area [51,57].

Currently, certain WT manufacturers incorporate fiber-optic sensors into the blades to reduce connections with the data logger and permit little to no weakening of the signal over a considerable distance. With the latest fiber optic sensing technologies, monitoring of stresses on the blades during rotation is easier and more accurate [27,75–77]. Kreuzer [73], Bang et al. [74] and Schroeder et al. [78] investigated the development of a high-speed-fiber Bragg-grating-based sensor array system for strain-based deflection shape estimation of WT structures.

### 12.3.1.1.5 Optical Fiber Monitoring (OFM)

OFM is growing as a reliable and cost-effective technique for WT SHM [71]. A network of sensors can be embedded in the blade structure to enable the measurement of five parameters that are critical to SHM. The five parameters include: (i) SM for monitoring the blade loading and vibration level; (ii) TM for likely over-heating; (iii) acceleration measurement for monitoring the pitch angle and rotor position; (iv) crack detection measurements; and (v) lightning detection for measuring the front steepness, maximum current, and specific energy [79–82].

The optical fibers must be mounted on the surface or embedded into the body of the monitored WT components. Therefore, OFM is complicated and expensive in real-world applications compared with other CM and fault detection methods [83,84]. However, due to technological progress, it is expected that the cost of OFM for WT SHM will decrease considerably in the future.

### 12.3.1.2. SUBSYSTEM CM TECHNIQUES BASED ON NDT

Malhotra et al. [54] defined NDT as "an examination, test, or evaluation performed on any type of test object without changing or altering it in any way". This is often done in order to determine the absence or presence of conditions or discontinuities that may have an effect on the usefulness or serviceability of the monitored object. NDTs may also be conducted to measure other tested object characteristics, i.e., size, dimension, configuration, or structures, including alloy content, hardness, and grain size. Nevertheless, these approaches are largely applied to localized areas. Thus, NDT technologies require more accurate prior knowledge of probable damage locations as well as the use of dedicated sensors [56].

### 12.3.1.2.1 VI

Based on human sensory capabilities, VI or observation is undoubtedly one of the oldest CMT and can serve as a supplement to other CMTs. VI includes the detection of sounds emitted by a functioning system, touch (temperature and vibration checking), and VI (e.g., deformation and aspects). This approach is generally used to monitor such components as rotor blades, nacelles, slip rings, yaw drives, bearings, generators, and transformers [53,85].

In several cases, VI is of great importance in identifying a problem that was not identified by other CMTs. Such cases may include loose parts, connections, terminals, and components; visibly worn or broken parts; excessive temperatures that reflect through the structure or housing, oil leakages, corrosion, chattering gears, or hot bearing housings [85–87]. Nevertheless, VI is limited to the identification of damages that are visible on the surface of a structure. Moreover, VI is labor intensive and highly subjective because the results depend on the experience and judgment of the inspector [88].

Today, the industry is implementing remote VI technologies to inspect gearboxes, WT blades, and other critical components [89]. AIT Inc. has developed a video boroscope or videoscope used to inspect the interior areas that are not accessible and can be efficient in revealing hairline cracks, corrosion, pitting, rubbing, and other defects [85]. Moreover, the Auto-Copter™ Corporation [87] has developed a flying remote VI device that enables inspection of WTs, thus increasing reliability and the number of daily inspections while eliminating the risk of personal injury.

### 12.3.1.2.2 Acoustic Emission (AE)

AE phenomena are based on the release of energy in the form of transitory elastic waves within a material via a dynamic deformation process [90]. Typically, sources of AE within a material are [91,92] crack initiation and propagation, breaking of fibers, and matrix cracking and fretting between surfaces at de-bonds or de-laminations. Unlike VA, AE can detect failures characterized by high-frequency vibrations ranging from 50 kHz to 1 MHz [93]. Piezoelectric transducers and optic fiber displacement sensors are often employed in this approach [94]. The most commonly measured AE parameters for diagnosis are amplitude, root mean square (RMS) value, energy, kurtosis, crest factor, counts, and events [95].

This method is typically applied for fault detection in gearboxes, bearings, shafts, and blades, and its advantages include a large frequency range and a relatively high signal-to-noise ratio. The main drawback of AE is its cost. Furthermore, only a few types of faults occur in the high-frequency range. Another limitation of AE is the attenuation of the signal during propagation. Therefore, an AE sensor must be located as close to its source as possible [96], which may pose a practical constraint in applying AE to certain wind machines.

Research was carried on the combined use of AE and VA data [97–99]. Soua et al. [99] presented the results of a combined vibration and AE monitoring effort that was performed over a continuous period of five years on an operating WT. Good results were obtained for the detection of defects, most notably in the gearbox, using special digital processing techniques, such as similarity analysis. Tan et al. [96] carried out a comparative experimental study on the diagnostic and prognostic capabilities of AE, VA, and spectrometric OA for spur gears and observed that based on the analysis of RMS levels, only the AE technique was more sensitive in detecting and monitoring faults than either the vibration or spectrometric OA.

### 12.3.1.2.3 UTTs

UTTs are extensively used by the wind energy industry for structural evaluation of WT towers and blades [27,92]. This method relies on elastic wave propagation and reflection within the material. Three different techniques can be used for this investigation: pulse-echo (Figure 11), through transmission, and pitch-catch [100,101]. Laser interferometric sensors, air-coupled transducers, electromagnetic acoustic transducers, or contact transducers are a subset of the sensors that can be used as the scanning sensor for acoustic wave field imaging, which is another UTT [72,102].

Implementation of UTTs implies one or more of the following measurements: time of flight or delay, path length, frequency, phase angle, amplitude, acoustic impedance, and angle of wave deflection [104]. Thus, signal-processing algorithms, including such time-frequency techniques as the Wigner-Ville distribution, Hilbert-Huang transform, and wavelet transform [100,105], can be used to extract additional information on internal defects.

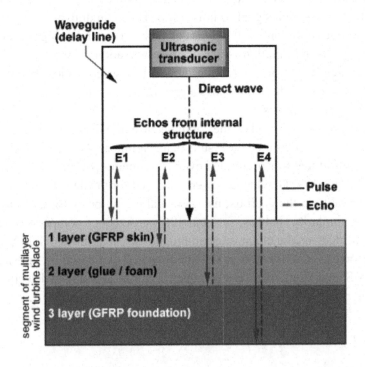

**FIGURE 11:** Principle of the pulse-echo technique used for the investigation of WT blades [100,103]. GFRP: glass fiber reinforced plastic. Reprinted/Reproduced with permission from [100]. Copyright 2008, Kaunas University of Technology.

Ultrasonic testing via wave propagation characteristics allows for the estimation of the location and nature of the detected failure. This approach provides a quick, reliable, and effective method for determining the material properties of the principal turbine components [106]. Ultrasound scanning allows personnel to see below the surface and check the laminate for dry glass fibers and de-lamination [107]. Unlike other NDT techniques (i.e., thermographic techniques), acoustic techniques are not as affected by temperature or air humidity [108].

### 12.3.1.2.4 Thermography Analysis (TA)

TA provides a wide range of diagnostic and monitoring applications in different equipment and machines, i.e., bearings, gear boxes, conveyor systems, drivers, motors, and electric generators. IR thermography is recognized as one of the most versatile and effective CM tools for use in the WT industry for control and diagnoses of electric parts and mechanical equipment [109,110]. This method is based on the fact that all working components emit heat and when a component in the system starts to malfunction, its temperature increases beyond the normal values [56]. IR temperature transmitters and high-resolution IR cameras are the sensors applied in TA, and results are typically interpreted visually [27,111].

Today, TA is primarily used for periodical manual inspection and can be used as a local or global technique because it is possible to assess the damage at the component or system level, depending on the resolution of the camera. However, TA is not appropriate for early fault detection because temperature develops slowly, as mentioned earlier [111,112]. Another important difficulty with TA for WTCM is that monitoring should be performed offline [111]. However, cameras and diagnostic software that are suitable for on-line process monitoring are currently entering the market [57].

### 12.3.1.2.5 Radiographic Inspection

Radiography (both film and digital) uses the well-known effects of an X-ray source on one side of a specimen and an X-ray-sensitive receptor on the other side. Although this method does provide useful information on the structural condition of the WT component under inspection, radiographic imaging using X-rays is rarely used in WECS industry [27]. The technique is highly efficient in detecting crack and de-lamination in the blade/rotor and tower structures.

### 12.3.1.2.6 Other ND WTCM Techniques

Other techniques are not widespread but are also used in the maintenance of WTs. In many cases, their performance is heavily influenced by the costs or excessive specialization, making them impractical in some situations. Examples are SMs in blades, voltage and current analysis, SPM, and magnetic flux leakage.

### 12.3.2 WT GLOBAL SYSTEM OR NONINTRUSIVE CMTS

Conventional subsystems CMTs (i.e., vibration, lubrication oil, and generator current signal analysis) require the deployment of a variety of sensors and computationally intensive analysis techniques [113]. The use of additional sensors and equipment increases costs and hardware complexity of the WECS. Furthermore, sensors and equipment are inevitably subject to failure, causing additional problems with system reliability and additional OM costs [114]. For these reasons, it is of interest to develop overall CMTs. These techniques are nonintrusive, low cost, and reliable.

Unlike subsystem CMTs, global systems CMTs enable the extraction of fault features with low calculation time from direct or indirect drives and fixed- or variable-speed WTs. In addition, these techniques can all be used in online and thus increase the WT reliability while reducing the downtime and OM costs [113–115]. Certain overall WTCM approaches include performance monitoring, power curve analyses, electrical signature, and supervisory control and data acquisition (SCADA) system data analysis.

## 12.3.2.1 PERFORMANCE MONITORING OR THE PROCESS PARAMETER TECHNIQUE

In WT performance monitoring, parameter readings of the capacity factors of the plant, power, wind velocity, rotor speed, and blade angle are compared with the values in operator manuals or manufacturer performance specifications to determine whether the system is performing at optimum efficiency. The relationships among power, wind velocity, rotor speed, and blade angle can be used for safeguarding purposes, and an alarm is generated in the case of large deviations. The detection margins are large to prevent false alarms [51,53]. Today, more intelligent usage of the signals based on parameter estimation and trending is not a common practice in the WT industry [51].

## 12.3.2.2 POWER SIGNAL ANALYSIS

Power quality is a high-interest area for WTCM because quality could degrade as a result of wind speed turbulence and switching events. From a global viewpoint, the mechanical power (torque times speed) measured on the WT drive shaft and the total three-phase electrical power measured from the terminals of the generator are the input and output of a WT system, respectively. Both energy flows are disturbed by WT abnormalities caused by mechanical or electrical faults [115]. Significant variations in the WT drive train torque are generally signs of abnormalities. Faults in the drive train cause either a torsional oscillation or shift in the T/$\omega$ ratio. By monitoring this ratio, certain fault conditions can be detected. For example, torque oscillations can be detected in a blade or rotor imbalance condition in the WT [71,116].

Peak power output, reactive power, voltage fluctuations, and harmonics greatly influence the power quality [117–120]. As an example, for a healthy WT, the output current is assumed to be sinusoidal:

$$i_H(t) = a\cos(\omega_i t) \tag{3}$$

A failure will cause a vibration in the shaft rotation at a certain frequency that can be detected by vibration sensors. The new shaft rotating speed is given by [121,122]:

$$\omega_F(t) = \omega_1 + c\cos(\omega_2 t) \qquad (4)$$

Therefore, the instantaneous phase for a faulty WT can be obtained:

$$\theta_F(t) = \int_0^t \omega_F(t)dt = \omega_1(t) + \gamma\sin(\omega_2 t) \qquad (5)$$

The current for the faulty WT can then be written as:

$$i_F(t) = a\cos[\omega_1 t + \gamma\sin(\omega_2 t)]$$
$$= a\cos(\omega_1 t)\cos(\gamma\sin(\omega_2 t)) - a\sin\,(\omega_1 t)\sin\,(\gamma\sin\,(\omega_2 t)) \qquad (6)$$

If we assume $c \ll \omega$, thus $\gamma \ll 1$. As a result, $\cos(\gamma\sin(\omega_2 t)) \approx 1$ and $\sin(\gamma\sin(\omega_2 t)) = \gamma\sin(\omega_2 t)$.

We will finally obtain:

$$i_F(t) = a\cos(\omega_1 t) - a\gamma\sin(\omega_1 t)\sin\,(\omega_2 t)$$
$$= a\cos(\omega_1 t) - \frac{a\gamma}{2}\cos((\omega_1 - \omega_2)t) + \frac{a\gamma}{2}\cos\,((\omega_1 - \omega_2)t) \qquad (7)$$

where $i_H(t)$ and $i_F(t)$ are the instantaneous currents for healthy and faulty WT, respectively; $\omega_1$ is the angular shaft rotation speed for a healthy WT; $\omega_2$ is the angular shaft rotation speed generated by the fault; $\omega_F$ is the shaft rotation speed for a faulty WT; a is the amplitude of the instantaneous current for a healthy WT; c is the amplitude of the current due to the WT fault; and $\gamma = c/\omega$. a, c and $\gamma$ are constants values. Frequency demodulation is used for feature extraction from Equation (7).

A mechanical failure can also lead to amplitude modulation of the output current. For a three-phase generator, the stator current $i_k(t)$ (k = 1,2,3) can be described in a discrete form as [121,123]:

$$i_k(n) = a_k(n) \cdot \cos(\omega n - \Phi_k) \tag{8}$$

where n = 0,...,N−1 is the sample index (N being the total number of received samples); and $\Phi_k = 2k\pi/3$ is the phase parameter. The angular frequency $\omega$ is equal to $2\pi f/F_e$, where f and $F_e$ are the supply and sampling frequencies, respectively. The amplitude $a_k(n)$ is related to the fault as follows:

- For a healthy WT, $a_k(n)$ is constant and there is no amplitude modulation;
- For a faulty WT, $a_k(n)$ is time variant and the current signal is modulated in amplitude.

Amplitude demodulation can be used for feature extraction using various techniques, such as the Concordia transform or Hilbert transform.

Wakui and Yokoyama [124] developed a sensorless wind-speed performance-monitoring method for stand-alone vertical-axis WTs using numerical analyses in a dynamic simulation model. Yang et al. [113,125,126] and Watson et al. [122] proposed a wind turbine condition monitoring technique (WTCMT) that uses the generator output power and rotational speed to derive a fault detection signal. The technique is based on a detection algorithm using a continuous-wavelet-transform-based adaptive filter to track the energy in the prescribed time-varying fault-related frequency bands in the power signal. A probabilistic model of the power curve based on copulas was developed by Gill et al. [127], for CM purposes. Copula analysis is likely to be useful in WTCM, particularly in early recognition of incipient faults, such as blade degradation, yaw, and pitch errors.

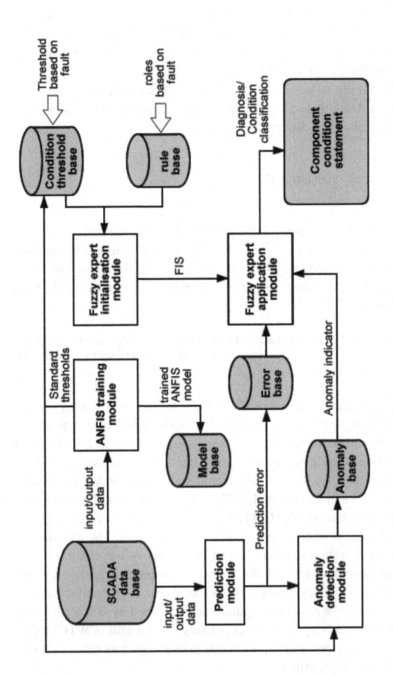

**FIGURE 12:** Overview of wind turbine condition monitoring (WTCM) based on supervisory control and data acquisition (SCADA) data analysis. FIS: fuzzy interference system.

### 12.3.2.3 WTCM BASED ON SIGNATURE ANALYSIS (SA)

SA is a simpler but more inclusive WTCM technique. Intensive research efforts have been focused on the use of SA to predict or detect electrical and mechanical faults in WECSs. Different signals can be detected (i.e., voltages, power, currents, or stray flux), and SA can be used to detect various faults (i.e., broken rotor bars, bearing failures, air gap eccentricity, and unbalanced rotors and blades) [50,128]. Yazidi et al. [129] proposed a monitoring system for WTs with double-fed induction generators based on stator and rotor current signatures. Different tests were performed in this work, and relevant results were obtained. The proposed CMS was efficient for detection of rotor and stator asymmetry in a double-fed induction machine connected to a back-to-back converter. A similar investigation was carried out by Douglas et al. [130].

Yang et al. [131] proposed a CMT based on both electrical and mechanical signatures. In addition to its versatile function, i.e., its ability to detect both mechanical and electrical faults, this technique removes the negative influence of variable wind in the machine CM. This work also investigated the possibility of detecting a WT mechanical fault (e.g., rotor imbalance fault and drive train mechanical fault) via power signal analysis.

### 12.3.2.4 WTCM BASED ON SCADA DATA ANALYSIS

In most modern WTs, SCADA systems are now common. The WTCM using SCADA data analysis is cost effective (data collection and sensor networks already in place) and reliable because it relies on the interpretation of SCADA data [132]. The SCADA system collects information extensively from key WT subassemblies using sensors fitted to the WT, i.e., anemometers, thermocouples, and switches. The operational data reflect either turbine status or measurements of signals, such as wind speed and direction, temperatures, currents, or pressures. This information can effectively reflect the real-time condition of a WECS, and by analyzing SCADA data, the relationship between different signals can be observed and the health of WT components deduced [133]. Neural networks and fuzzy logic are other examples of the common tools for data analysis. An overview of WTCM based on SCADA data analysis is presented in Figure 12.

Several recent studies on SCADA data for WECS CM can be found in the literatures [132,134–137]. A wind turbine condition monitoring system (WTCMS) based on SCADA using normal behavior models and fuzzy logic was presented in [135]. This CMS is designed to detect trends and patterns in SCADA data and predict possible failures. Another recent research study by Li et al. [136] focused on improving the fuzzy synthetic condition assessment of a WT generator system. The results indicated that the evaluation of dynamic limits and deterioration degree functions for the characteristic variables for WECSs could be improved by analyzing SCADA data with the improved fuzzy synthetic model.

However, the WT SCADA system does not collect all of the information necessary to conduct a full CM of a WT because it was not initially designed for CM purposes. Furthermore, although SCADA techniques are widely applied to WT, the data rate of once every 5–10 min is too slow for most rotating machine fault diagnoses [116,131,133]. Another concern is that the values of SCADA data vary over wide ranges under varying operational conditions, and it is difficult to detect an incipient fault from raw SCADA data without an appropriate data analysis tool [137].

This section provides a status update on different methods and techniques used for WTCM. Table 1 presents an overview on the state of the art for WTCMSs, including possible failures and corresponding monitoring techniques for various WT components and subsystems. The following section focuses on new trends in WTCM with respect to the wind industry's evolution, and implications for challenges in the research area will be discussed based on these tendencies.

## 12.4 NEW TRENDS AND FUTURE CHALLENGES IN WTCMTS

The state-of-the-art maintenance strategy in the wind machine industry is defined by the implementation of on-line continuous CMS. The authors of [133,138] conducted respective surveys of: (i) commercially available CMSs for WT; and (ii) commercially available SCADA data analysis tools for WT health monitoring. The survey in [138] elaborated on the methods used by 20 suppliers and concluded that nearly all of them focus on the same subassemblies, i.e., blades, main bearings, gearbox internals, gearbox bearings, and generator bearings.

Furthermore, VM, OM, and fiber optic monitoring are the most frequently used monitoring techniques. The study in [133] addressed 17 SCADA data analysis tools for WTCM. Among the 17 products, three were developed by WT manufacturers, two by renewable energy consultancies, up to nine by industrial software companies, two by an electrical equipment provider, and only one by a WT operating company.

## 12.4.1 NEW TRENDS IN WTCMSS

The current trend in the wind energy industry is the use of larger WTs in remote locations, which are increasingly situated offshore for optimal wind conditions. Both the size and location factors have led to maintenance challenges that are unique compared with those of traditional power generation systems [66]. To cope with this reality, WTCMS manufacturers must improve the existing monitoring techniques and/or develop more appropriate techniques. The future goal in CMS is to continue to minimize the efforts required from operators through the use of intelligent software algorithms and automated analysis.

The WT industry is moving toward intelligent machine health management (IMHM), which is a fourth-generation maintenance strategy. The final objective is to provide WECSs that are capable of understanding and making decisions without human intervention. This goal implies the use of intelligent condition-based maintenance systems based on RCM mechanisms. Thus, the following tendencies can be mentioned with respect to the new tendencies in the WTCM industry [52].

### 12.4.1.1 TOWARD SMART MONITORING

The purpose of this effort is to develop a CMS that is self-contained. Such systems could be operated by trained personnel but would not require specialists for the interpretation of results because a smart monitoring system will be able to perform classification and prediction operations [139]. Therefore, the number of turbines that a technician is able to oversee might double. Moreover, smart WTCMSs will integrate built-in hardware auto-diagnostics that continuously check all sensors, cabling, and electronics for any faults, signal interruption, shorts, or power failures. Any malfunctions trigger an alarm. Indeed, false warnings and false alarms occur on a regular basis with actual CMSs [119,140–143]. The use of smart monitoring will aid in avoiding such situations.

Automation of CM and diagnostic systems will also be an important development as WT operators acquire a larger number of turbines and manual inspection of data becomes impractical. Furthermore, it is essential that methods for reliable, automatic diagnosis are developed with consideration of multiple signals to improve detection and increase operator confidence in alarm signals [25,138].

## 12.4.1.2 NECESSITY OF REMOTE AND E-MONITORING

Considering: (i) the tendency toward the use of offshore WTs; and (ii) the fact that wind parks are geographically dispersed and often located in remote areas, cost considerations make it necessary to reevaluate the traditional monitoring setup. Thus, remote CM of WECSs is gaining popularity in the industry and can be implemented as either standalone or networked systems. Remote CM involves monitoring the condition of a component at a location far away from the immediate vicinity of the component in question. E-monitoring and CM using the Internet improves remote monitoring by providing worldwide remote capabilities. Because browsers reside on many platforms, internet-based CMSs can be accessed by multiple users working on any type of operating system [144–146]. In short, wireless technologies will help to optimize the cost and efficiency of WTCMSs.

## 12.4.1.3 IN-SERVICE SHM

Given the increased size of modern turbines and their growing cost and fabrication sophistication, i.e., high-tech, complex, and constructed with composite materials, SHM is becoming increasingly important to both operators and insurers [143]. The necessity for continuous in-service SHM is a reality because these complex structures are fragile. For example, if any blade fails, the rotor can become unbalanced and might lead to the destruction of the entire WT [147]. Therefore, it is important to acquire early indications of structural or mechanical problems that will allow operators to better plan for maintenance, possible operation of the machine in a derated condition rather than taking the turbine off-line, or, in the case of an emergency, shutdown of the machine to avoid further damage.

The development of real-time, remote, wireless, and smart SHM is playing an increasingly important role. Such monitoring systems designed for the continuous assessment of structural performance and safety should be comprehensive and include functions for self-diagnostics and management of the SHM system [88,148,149]. Similar SHM techniques are already used in certain industries, such as aeronautics, where they are applied for the SHM of aircraft composite structures [150,151]. Additionally, there is a tendency to require ambient energy harvesting for powering wireless sensors [152,153]. However, a major limitation in the field of energy harvesting is the fact that the energy generated by harvesting devices is far too small to directly power most electronics. Therefore, (i) efficient, innovative, and adapted methods of storing electric energy; and (ii) more energy-efficient sensors are the key technologies that will allow energy harvesting to become a source of power for electronics and wireless sensors [154–156].

Two different approaches are emerging in the field of WTSHM. The first and more practical approach is the development of appropriate non-contact and remote NDT/inspecting technologies for in-service WTSHM because non-contact and remote NDTs have overwhelming advantages in terms of on-line testing and inspection. The second approach consists of equipping the WT with a SHM system consisting of a network of sensors, data acquisition units, and an on-site server installed in the WT maintenance room. The sensors (accelerometers, displacement transducers, and temperature sensors) are placed at different levels inside and outside the steel tower and on the foundation of the WT. In this last case, microchip path antennas are increasingly used for sensing, ambient energy harvesting, and data transmission [157–160].

## 12.4.1.4 INTEGRATION AND INTERACTION OF MONITORING AND CONTROL SYSTEMS

Today's standard CMSs essentially still operate in stand-alone mode, i.e., independent of the WT controller. The CMSs are increasingly integrated with control functions and included in maintenance concepts [161]. The full integration of CM capabilities within the WT control system is beneficial with regard to three different aspects: (i) cost benefits; (ii) technical benefits; and (iii) quality benefits. An overview of the benefits of controller-integrated CMSs was presented in [162,163] and can be summarized as follows:

Cost benefits:

- Lower hardware costs due to industrial mass production and fewer components;
- Lower installation and cabling costs due to integration in the existing control cabinet and communication with the main controller via bus systems;
- Fewer required parts because no additional voltage transformers, communication modules, uninterruptible power supply (UPS), or similar devices are needed;
- Reduced analysis because fewer false alarms occur.
- Technical benefits:
- No measurement if interference signals are present;
- Higher-quality raw data for analysis;
- Fewer false alarms;
- Reduced scatter leads to improved fault detection;
- Integration of further signals (e.g., temperature, pressure, and current) enables integrated signal/system monitoring.
- Quality benefits:
- Reliable hardware from established industrial suppliers;
- Mass production with high-quality standards.

## 12.4.1.5 ESTIMATION OF THE REMAINING COMPONENT LIFE SERVICE

The limited accessibility of offshore wind farms requires new maintenance and repair strategies. In fact, offshore wind farms are likely to be unreachable for several months out of a year, especially if sited in the

North Sea and polar regions [164,165]. Thus, maintenance and repair activities must be carried out during seasons in which the turbines are accessible. Components that are likely to fail during periods of inaccessibility must be replaced. This approach is referred to as a "condition-dependent and predictive maintenance and repair strategy" [38,166]. Such a strategy requires comprehensive knowledge of the actual condition and the remaining lifetime of the turbine components. Such knowledge can be provided by CMSs. For those components, a count of their lifetime fatigue load can provide information on the condition and remaining lifetime. However, current CMSs are not able to assure that a given component will not fail, nor can they prevent a failure.

## 12.4.2 FUTURE RESEARCH CHALLENGES IN WTCMTS

Although CM technologies face various challenges in WT applications, they are still necessary and valuable. As with any technology, there is room for improvement such that these systems can be better utilized to benefit the wind industry. Based on the provided discussion on new trends in WTCM, selected key points that must be addressed by further research are listed as follows:

- Determine the most cost-effective measurement or monitoring strategy.
- Automate the "experts" in data interpretation to automate actionable recommendations.
- Develop reliable and accurate prognostic techniques.
- Improve the use of SCADA system data (normally only stored at 10-min intervals) to provide a more reliable, flexible, and efficient tool for automatic WT monitoring and control [133].
- Develop smart, wireless, and energy-efficient sensors that will offer opportunities for placing sensors in difficult-to-reach locations, electrically noisy environments, and mobile applications in which wires cannot be installed.
- Focus on providing the newest and industry-proven signal processing algorithms for extracting the key features of a signal to predict machine component health.
- Combine numerical simulation analysis with testing, inspecting, and monitoring technologies. The finite element method is one such interesting tool that has traditionally been used in the development of WT blades, primarily to investigate the global behavior in terms of eigenfrequencies, tip deflections, and global stress/strain levels [167]. An advantage of using the finite

element method is that complex load cases that represent actual wind conditions can be analyzed once the model is set up and calibrated. Moreover, this method will considerably reduce the cost of testing, inspecting, and monitoring for WTs, especially SHM.
• Develop innovative, adapted, and efficient methods of harvesting and storing electric energy for autonomous and wireless sensors.

Other technological advances that must be developed in WTCMs include advancements in diagnostic and prognostic software, acceptance of communication protocols, and developments in maintenance software applications and computer networking technologies [146]. Although these future research areas may appear challenging to address, they also represent great opportunities for CM to boost the success of the wind industry by reducing the COE and increasing its competitiveness.

## 12.5 CONCLUSIONS

WT technology has greatly advanced in a relatively short time span. Among the technologies successfully transferred from applications in other industries, CMSs enable early detection and diagnosis of potential component failures and serve as a platform for implementing CM practices.

This paper performed an inventory and classification of WTCMTs and has highlighted the fact that a combination of preventive and reactive maintenance strategies can improve reliability, availability, and maintainability of WTs while reducing maintenance costs. An overview of CM and the maintenance process in the WT industry enabled the presentation of a global diagram linking the various concepts, and a comprehensive review of WTCM techniques and methods was carried out.

For new trends in WTCM, the wind energy industry's tendency to use larger WTs in remote locations implies the need for remote, intelligent, and integrated CMSs. In particular, efforts should be directed toward improving the capacity of CMSs for failure prognostics and determination of remaining equipment life. Finally, this work addressed certain important and challenging areas of research that should be explored for the industry to better cope with the major innovations that are likely to occur in the WTCM industry.

The authors would like to thank the Natural Science and Engineering Research Council of Canada (NSERC) for financially supporting this research. Authors' gratefulness also goes to the editor and four anonymous reviewers for their valuable comments and suggestions that appreciably improved the quality of this paper.

## REFERENCES

1. Fried, L.; Sawyer, S.; Shukla, S.; Qiao, L. Global Wind Report 2012—Annual Market Update; Global Wind Energy Council (GWEC): Brussels, Belgium, 2013. Available online: http://www.gwec.net (accessed on 22 February 2014).
2. Zhang, P. Small Wind World Report 2012; World Wind Energy Association (WWEA): Bonn, Germany, 2012.
3. Renewables 2012: Global Status Report; REN21: Paris, France, 2012.
4. Ribrant, J. Reliability Performance and Maintenance—A Survey of Failures in Wind Power Systems. Master's Thesis, School of Electrical Engineeringm, KTH Royal Institute of Technology, Stockholm, Sweden.
5. Hines, V.; Ogilvie, A.; Bond, C. Continuous Reliability Enhancement for Wind (CREW) Database: Wind Turbine Reliability Benchmark Report; Sandia National Laboratories: Albuquerque, NM, USA, 2013.
6. Walford, C.A. Wind Turbine Reliability: Understanding and Minimizing Wind Turbine Operation and Maintenance Costs. Sandia Report No. SAND2006-1100; Sandia National Laboratories: Albuquerque, NM, USA, 2006.
7. Tchakoua, P.; Wamkeue, R.; Tameghe, T.A.; Ekemb, G. A Review of Concepts and Methods for Wind Turbines Condition Monitoring. Proceedings of the 2013 World Congress on Computer and Information Technology (WCCIT), Sousse, Tunisia, 22–24 June 2013; pp. 1–9.
8. Fischer, K.; Besnard, F.; Bertling, L. Reliability-centered maintenance for wind turbines based on statistical analysis and practical experience. IEEE Trans. Energy Convers. 2012, 27, 184–195.
9. McMillan, D.; Ault, G. Condition monitoring benefit for onshore wind turbines: Sensitivity to operational parameters. IET Renew. Power Gener. 2008, 2, 60–72.
10. Amayri, A.; Tian, Z.; Jin, T. Condition Based Maintenance of Wind Turbine Systems Considering Different Turbine Types. Proceedings of the 2011 International Conference on Quality, Reliability, Risk, Maintenance, and Safety Engineering (ICQR2MSE), Xi'an, Shannxi, China, 17–19 June 2011; pp. 596–600.
11. Besnard, F.; Bertling, L. An approach for condition-based maintenance optimization applied to wind turbine blades. IEEE Trans. Sustain. Energy 2010, 1, 77–83.
12. Echavarria, E.; Hahn, B.; van Bussel, G.J.; Tomiyama, T. Reliability of wind turbine technology through time. J. Sol. Energy Eng. 2008, 130, doi:10.1115/1.2936235.
13. Hahn, B.; Durstewitz, M.; Rohrig, K. Reliability of Wind Turbines. In Wind Energy; Springer: Berlin/Heidelberg, Germany, 2007; pp. 329–332.

14. Ahmad, R.; Kamaruddin, S. An overview of time-based and condition-based maintenance in industrial application. Comput. Ind. Eng. 2012, 63, 135–149.

15. Andrawus, J.A.; Watson, J.; Kishk, M. Wind turbine maintenance optimisation: Principles of quantitative maintenance optimisation. Wind Eng. 2007, 31, 101–110.

16. Andrawus, J.A. Maintenance Optimisation for Wind Turbines. Ph.D. Thesis, Robert Gordon University, Aberdeen, UK, April 2008. Hahn, B. Reliability Assessment of Wind Turbines in Germany. Proceedings of the 1999 Europeen Wind Energy Conference, Nice, France, 1–5 March 1999.

17. Tavner, P.J.; Xiang, J.; Spinato, F. Reliability analysis for wind turbines. Wind Energy 2007, 10, 1–18.

18. Guo, H.; Watson, S.; Tavner, P.; Xiang, J. Reliability analysis for wind turbines with incomplete failure data collected from after the date of initial installation. Reliab. Eng. Syst. Saf. 2009, 94, 1057–1063.

19. Spinato, F.; Tavner, P.J.; van Bussel, G.J.W.; Koutoulakos, E. Reliability of wind turbine subassemblies. IET Renew. Power Gener. 2008, 3, 387–401.

20. Amirat, Y.; Benbouzid, M.E.H.; Member, S.; Bensaker, B.; Wamkeue, R. Condition Monitoring and Fault Diagnosis in Wind Energy Conversion Systems: A Review. Proceedings of the IEEE International Electric Machines & Drives Conference, IEMDC '07, Antalya, Turkey, 3–5 May 2007; Volume 2, pp. 1434–1439.

21. Ribrant, J.; Bertling, L. Survey of Failures in Wind Power Systems with Focus on Swedish Wind Power Plants during 1997–2005. Proceedings of the 2007 IEEE Power Engineering Society General Meeting, Tampa, FL, USA, 24–28 June 2007; pp. 1–8.

22. Villa, L.F.; Renones, A.; Pera, J.R.; de Miguel, L.J. Statistical fault diagnosis based on vibration analysis for gear test-bench under non-stationary conditions of speed and load. Mech. Syst. Signal Process. 2012, 29, 436–446.

23. Crabtree, C.J.; Feng, Y.; Tavner, P.J. Detecting Incipient Wind Turbine Gearbox Failure: A Signal Analysis Method for On-line Condition Monotoring. Proceedings of European Wind Energy Conference (EWEC 2010), Warsaw, Poland, 20–23 April 2010; pp. 154–156.

24. Crabtree, C.J. Condition Monitoring Techniques for Wind Turbines. Ph.D. Thesis, Durham University, Durham, UK, February 2011.

25. Aziz, M.A.; Noura, H.; Fardoun, A. General Review of Fault Diagnostic in Wind Turbines. Proceedings of the 2010 18th Mediterranean Conference on Control & Automation (MED), Marrakech, Morocco, 23–25 June 2010; pp. 1302–1307.

26. García, F.P.; Tobias, A.M.; Pinar, J.M.; Papaelias, M. Condition monitoring of wind turbines: Techniques and methods. Renew. Energy 2012, 46, 169–178.

27. Hameed, Z.; Hong, Y.S.; Cho, Y.M.; Ahn, S.H.; Song, C.K. Condition monitoring and fault detection of wind turbines and related algorithms: A review. Renew. Sustain. Energy Rev. 2009, 13, 1–39.

28. Lu, B.; Li, Y.; Wu, X.; Yang, Z. A Review of Recent Advances in Wind Turbine Condition Monitoring and Fault Diagnosis. Proceedings of the IEEE Power Electronics and Machines in Wind Applications, PEMWA 2009, Lincoln, NE, USA, 24–26 June 2009; pp. 1–7.

29. Orsagh, R.F.; Lee, H.; Watson, M.; Byington, C.S.; Power, J. Advance Vibration Monitoring for Wind turbine Health Management; Impact Technologies, LLC: Rochester, NY, USA, 2006.

30. Dhillon, B.S. Engineering Maintenance: A Modern Approach; CRC Press: Boca Raton, FL, USA, 2002.

31. Madsen, B.N. Condition Monitoring of Wind Turbines by Electric Signature Analysis. Master's Thesis, Technical University of Denemark, Copenhagen, Denmark, October 2011.

32. Wang, L.; Yeh, T.-H.; Lee, W.-J.; Chen, Z. Benefit evaluation of wind turbine generators in wind farms using capacity-factor analysis and economic-cost methods. IEEE Trans. Power Syst. 2009, 24, 692–704.

33. Al-Ahmar, E.; Hachemi, M.E.; Turri, S. Wind energy conversion systems fault diagnosis using wavelet analysis. Int. Rev. Electr. Eng. 2008, 3, 646–652.

34. De Oliveria, W.S.; Fernandes, A.J. Cost-effectiveness analysis for wind energy projects. Int. J. Energy Sci. 2012, 2, 15–21.

35. Byon, E.; Ntaimo, L.; Ding, Y. Optimal maintenance strategies for wind turbine systems under stochastic weather conditions. IEEE Trans. Reliab. 2010, 59, 393–404.

36. Blischke, W.R.; Karim, M.R.; Prabhakar Murthy, D.N. Warranty Data Collection and Analysis; Springer: London, UK, 2011.

37. Nilsson, J.; Bertling, L. Maintenance management of wind power systems using condition monitoring systems—Life cycle cost analysis for two case studies. IEEE Trans. Energy Convers. 2007, 22, 223–229.

38. Hameed, Z.; Ahn, S.H.; Cho, Y.M. Practical aspects of a condition monitoring system for a wind turbine with emphasis on its design, system architecture, testing and installation. Renew. Energy 2010, 35, 879–884.

39. Wiggelinkhuizen, E.J.; Verbruggen, T.W.; Braam, H.; Rademakers, L.W.M. M.; Xiang, J.; Watson, S.; Giebel, G.; Norton, E.; Tipluica, M.C.; MacLean, A.; et al. CONMOW: Condition Monitoring for Offshore Wind Farms. Proceedings of the European Wind Energy Conference (EWEC2007), Milan, Itlay, 7–10 May 2007.

40. Renewable Energy Technologies: Cost Analysis Series; The International Renewable Energy Agency (IRENA): Abu Dhabi, United Arab Emirates, 2012; Volume 1.

41. Arunraj, N.S.; Maiti, J. Risk-based maintenance—Techniques and applications. J. Hazard. Mater. 2007, 142, 653–661.

42. Sheng, C.; Li, Z.; Qin, L.; Guo, Z.; Zhang, Y. Recent progress on mechanical condition monitoring and fault diagnosis. Procedia Eng. 2011, 15, 142–146.

43. Jardine, A.K.S.; Lin, D.; Banjevic, D. A review on machinery diagnostics and prognostics implementing condition-based maintenance. Mech. Syst. Signal Process. 2006, 20, 1483–1510.

44. Utne, I.B.; Brurok, T.; Rodseth, H. A structured approach to improved condition monitoring. J. Loss Prev. Process Ind. 2012, 25, 148–188.

45. Bengsston, M. On Condition Based Maintenance ant Its Implimentation in Industrial Settings. Ph.D. Thesis, Mälardalen University, Västerås, Sweden, 2007.

46. Bengtsson, M.; Olsson, E.; Funk, P.; Jackson, M. Technical Design of Condition Based Maintenance System—A Case Study using Sound Analysis and Case-Based Reasoning. Proceedings of the 8th Congress on Maintenance and Reliability Conference, MARCON 2004, Knoxville, TN, USA, 2–5 May 2004.

47. Simeón, E.A.; Álvares, A.J. An Expert System for Fault Diagnostics in Condition Based Maintenance. ABCM Symp. Ser. Mechatron. 2010, 4, 304–313.

48. Maintenance Terminology. SS-EN 13306; Swedish Standards Institute: Stockholm, Sweden, 2011.

49. Popa, L.M.; Jensen, B.-B.; Ritchie, E.; Boldea, I. Condition Monitoring of Wind Generators. Proceedings of the 38th IAS Annual Meeting, Conference Record of the Industry Applications Conference, Salt Lake City, UT, USA, 12–16 October 2003; Volume 3, pp. 1839–1846.

50. Verbruggen, T. Wind Turbine Operation & Maintenance based on Condition Monitoring WT-Ω. ECN-C-03-047; Energieonderzoek Centrum Nederland (ECN): Petten, The Netherlands, 2003.

51. Tchakoua, P.; Wamkeue, R.; Slaoui-Hasnaoui, F.; Tameghe, T.A.; Ekemb, G. New Trends and Future Challenges for Wind Turbines Condition. Proceedings of the 2013 International Conference on Control Automation and Information Sciences (ICCAIS), Nha Trang, Vietnam, 25–28 November 2013; pp. 238–245.

52. Saeed, A. Online Condition Monitoring System for Wind Turbine. Master's Thesis, Blekinge Institute of Technology, Karlskrona, Sweden, 2008.

53. Malhotra, V.M.; Carino, N.J. Handbook on Nondestructive Testing of Concrete; CRC Press: Boca Raton, FL, USA, 2004.

54. Gamidi, S.H. Non Destructive Testing of Structures. Master's Thesis, Indian Institute of Technology, Bombay, Libya, November 2009.

55. Hellier, C.J. Handbook of Nondestructive Evaluation; McGraw-Hill Professional Publishing: New York, NY, USA, 2003.

56. Elforjani, M.A. Condition Monitoring of Slow Speed Rotating Machinery Using Acoustic Emission Technology. Ph.D. Thesis, Cranfield University, Cranfield, UK, June 2010.

57. Khan, M.M.; Iqbal, M.T.; Khan, F. Reliability and Condition Monitoring of a Wind Turbine. Proceedings of the 2005 Canadian Conference on Electrical and Computer Engineering, Saskatoon, SK, Canada, 1–4 May 2005; pp. 1978–1981.

58. Jablonsky, A.; Barszcz, T.; Bielecka, M. Automatic validation of vibration signals in wind farm distributed monitoring systems. Measurement 2011, 44, 1954–1967.

59. Mechanical Vibration—Evaluation of Machine Vibration by Measurements on Non-Rotating Parts—Part 1: General Guidelines. ISO 10816-1:1995; International Organization for Standardization (ISO): Geneva, Switzerland, 1995.

60. Barrett, M.P.; Stover, J. Understanding Oil Analysis: How It Can Improve the Reliability of Wind Turbine Gearboxes; Gear Technology: Elk Grove Village, IL, USA, 2013; pp. 104–111.

61. Dupuis, R. Application of Oil Debris Monitoring For Wind Turbine Gearbox Prognostics and Health Management. Proceedings of the Annual Conference of the Prognostics and Health Management Society 2010, Portland, OR, USA, 10–16 October 2010.

62. Goncalves, A.C.; Campos, J.B. Predictive maintenance of a reducer with contaminated oil under an excentrical load through vibration and oil analysis. J. Braz. Soc. Mech. Sci. Eng. 2011, 33, 1–7.

63. Hamilton, A.; Quail, F. Detailed state of the art review for the different on-line/in-line oil analysis techniques in context of wind turbine gearboxes. J. Tribol. 2011, 133, doi:10.1115/1.4004903.

64. Walford, C.; Roberts, D. Condition Monitoring of Wind Turbines: Technology Overview, Seeded-Fault Testing, and Cost-Benefit Analysis. Technical Report 1010149; Global Energy Concepts, LLC: Kirkland, WA, USA, 2006.

65. Sheng, S.; Veers, P. Wind Turbine Drivetrain Condition Monitoring—An Overview, Mechanical Failures Prevention Group: Applied Systems Health Management Conference 2011, Virginia Beach, VA, USA, 10–12 May 2011.

66. Du, L.; Zhe, J. A high throughput inductive pulse sensor for online oil debris monitoring. Tribol. Int. 2011, 44, 175–179.

67. Hasan, I.; Rahaman, M.I. Intelligent Diagnostics and Predictive Maintenance Sensor System for Electrical Fault Diagnosis of Wind Turbine System. Proceedings of the Global Engineering, Science and Technology Conference 2012, Dhaka, Bangladesh, 28–29 December 2012.

68. Park, J.-Y.; Lee, J.-K.; Oh, K.-Y.; Lee, J.-S.; Kim, B.-J. Design of Simulator for 3MW Wind Turbine and its Condition Monitoring System. Proceedings of the International MultiConference of Engineers and Computer Scientists, IMECS 2010, Kowloon, Hong Kong; Volume II, pp. 930–933. 17–19 March 2010.

69. Jayaswal, P.; Wadhwani, A.; Mulchandani, K. Machine fault signature analysis. Int. J. Rotat. Mach. 2008, 2008, doi:10.1155/2008/583982.

70. Gong, X. Online Nonintrusive Condition Monitoring and Fault Detection for Wind Turbines. Ph.D. Thesis, Department of Electrical Engineering, University of Nebraska–Lincoln, Lincoln, NE, USA, August 2012.

71. Ciang, C.C.; Lee, J.-R.; Bang, H.-J. Structural health monitoring for a wind turbine system: A review of damage detection methods. Meas. Sci. Technol. 2008, 19, doi:10.1088/0957-0233/19/12/122001.

72. Kreuzer, M. Strain Measurement with Fiber Bragg Grating Sensors. S2338-1.0 e; HBM GmbH: Darmstadt, Germany, 2006.

73. Bang, H.-J.; Kim, H.-I.; Lee, K.-S. Measurement of strain and bending deflection of a wind turbine tower using arrayed FBG sensors. Int. J. Precis. Eng. Manuf. 2012, 13, 2121–2126.

74. Lee, J.M.; Hwang, Y. A novel online rotor condition monitoring system using fiber Bragg grating (FBG) sensors and a rotary optical coupler. Meas. Sci. Technol. 2008, 19, doi:10.1088/0957-0233/19/6/065303.

75. Kang, L.-H.; Kim, D.-K.; Han, J.-H. Estimation of dynamic structural displacements using fiber Bragg grating strain sensors. J. Sound Vib. 2007, 305, 534–542.

76. Turner, A.; Graver, T.W. Structural monitoring of wind turbine blades using fiber optic Bragg grating strain sensors. Exp. Mech. Emerg. Energy Syst. Mater. 2011, 5, 149–154.

77. Schroeder, K.; Ecke, W.; Apitz, J.; Lembke, E.; Lenschow, G. A fibre Bragg grating sensor system monitors operational load in a wind turbine rotor blade. Meas. Sci. Technol. 2006, 17, doi:10.1088/0957-0233/17/5/S39.

78. Kramer, S.G.; Leon, F.P.; Appert, B. Fiber Optic Sensor Network for Lightning Impact Localization and Classification in Wind Turbines. Proceedings of the 2006

IEEE International Conference on Multisensor Fusion and Integration for Intelligent Systems, Heidelberg, Germany, 3–6 September 2006; pp. 173–178.

79. Rademakers, L.W.M.M.; Vebruggen, T.W.; van der Werff, P.A.; Korterink, H.; Richon, D.; Rey, P.; Lancon, F. Fiber Optic Blade Monitoring. Proceedings of the European Wind Energy Conference, London, UK, 22–25 November 2004; pp. 22–25.

80. Eum, S.; Kageyama, K.; Murayama, H.; Uzawa, K.; Ohsawa, I.; Kanai, M.; Igawa, H. Process/Health Monitoring for Wind Turbine Blade by Using FBG Sensors with Multiplexing Techniques. Proceedings of the 19th International Society for Optics and Photonics, Perth, Australia, 14–18 April 2008; Volume 7004.

81. Shin, C.; Chen, B.; Cheng, J.; Liaw, S. Impact response of a wind turbine blade measured by distributed FBG sensors. Mater. Manuf. Process. 2010, 25, 268–271.

82. Guemes, A.; Fernandez-Lopez, A.; Soller, B. Optical fiber distributed sensing-physical principles and applications. Struct. Health Monit. 2010, 9, 233–245.

83. Merzbacher, C.; Kersey, A.; Friebele, E. Fiber optic sensors in concrete structures: A review. Smart Mater. Struct. 1996, 5, doi:10.1088/0964-1726/5/2/008.

84. Moragues Pons, J. Practical Experiments on the Efficiency of the Remote Presence: Remote Inspection in an Offshore Wind Turbine. Master's Thesis, Norwegian University of Science and Technology, Trondheim, Norway, 21 June 2012.

85. Sheppard, R.; Puskar, F.; Waldhart, C. SS: Offshore Wind Energy Special Session: Inspection Guidance for Offshore Wind Turbine Facilities. Proceedings of the Offshore Technology Conference, Houston, TA, USA, 3–6 May 2010.

86. Effren, D. Automated Turbine Inspection; AutoCopter™ Corporation: Charlotte, NC, USA, 2011.

87. Smarsly, K.; Law, K.H. Advanced Structural Health Monitoring Based on Multi-Agent Technology. In Computation for Humanity: Information Technology to Advance Society; Taylor & Francis Group, 2012.

88. McGugan, M.; Larsen, G.C.; Sorensen, B.F.; Borum, K.K.; Engelhardt, J. Fundamentals for Remote Condition Monitoring of Offshore Wind Turbines; Danmarks Tekniske Universitet, Risø Nationallaboratoriet for Bæredygtig Energi: Risø, Denmark, 2008.

89. Balageas, D.; Fritzen, C.-P.; Güemes, A. Structural Health Monitoring; ISTE Ltd.: London, UK, 2006.

90. Schubel, P.; Crossley, R.; Boateng, E.; Hutchinson, J. Review of structural health and cure monitoring techniques for large wind turbine blades. Renew. Energy 2013, 51, 113–123.

91. Hyers, R.; McGowan, J.; Sullivan, K.; Manwell, J.; Syrett, B. Condition monitoring and prognosis of utility scale wind turbines. Energy Mater. 2006, 1, 187–203.

92. Watson, M.; Sheldon, J.; Amin, S.; Lee, H.; Byington, C.; Begin, M. A Comprehensive High Frequency Vibration Monitoring System for Incipient Fault Detection and Isolation of Gears, Bearings and Shafts/Coupl. Proceedings of the ASME Turbo Expo 2007: Power for Land, Sea, and Air, Montreal, QC, Canada, 14–17 May 2007; Volume 5, pp. 885–894.

93. Lading, L.; McGugan, M.; Sendrup, P.; Rheinlander, J.; Rusborg, J. Fundamentals for Remote Structural Health Monitoring of Wind Turbine Blades-A Preproject,

Annex B: Sensors and Non-Destructive Testing Methods for Damage Detection in Wind Turbine Blades; Riso National Laboratory: Roskilde, Denmark, 2002.

94. Mba, D.; Rao, R.B. Development of acoustic emission technology for condition monitoring and diagnosis of rotating machines; bearings, pumps, gearboxes, engines and rotating structures. Shock Vib. Dig. 2006, 38, 3–16.

95. Tan, C.K.; Irving, P.; Mba, D. A comparative experimental study on the diagnostic and prognostic capabilities of acoustics emission, vibration and spectrometric oil analysis for spur gears. Mech. Syst. Signal Process. 2007, 21, 208–233.

96. Loutas, T.; Kalaitzoglou, J.; Sotiriades, G.; Kostopoulos, V. The Combined Use Of Vibration, Acoustic Emission And Oil Debris Sensor Monitored Data Coming From Rotating Machinery For The Development Of A Robust Health Monitoring System. Availble online: http://maritime-conferences.com/asranet2010-conference/asranet2008/53%20Kostopoulos,%20V.pdf (accessed on 22 February 2011).

97. Loutas, T.; Sotiriades, G.; Kalaitzoglou, I.; Kostopoulos, V. Condition monitoring of a single-stage gearbox with artificially induced gear cracks utilizing on-line vibration and acoustic emission measurements. Appl. Acoust. 2009, 70, 1148–1159.

98. Soua, S.; Lieshout, P.V.; Perera, A.; Gan, T.-H.; Bridge, B. Determination of the combined vibrational and acoustic emission signature of a wind turbine gearbox and generator shaft in service as a pre-requisite for effective condition monitoring. Renew. Energy 2013, 51, 175–181.

99. Raisutis, R.; Jasiuniene, E.; Sliteris, R.; Vladisauskas, A. The review of non-destructive testing techniques suitable for inspection of the wind turbine blades. Ultrasound 2008, 63, 26–30.

100. Rose, J.L. Ultrasonic guided waves in structural health monitoring. Key Eng. Mater. 2004, 270, 14–21.

101. Lee, J.-R.; Takatsubo, J.; Toyama, N.; Kang, D.-H. Health monitoring of complex curved structures using an ultrasonic wavefield propagation imaging system. Meas. Sci. Technol. 2007, 18, doi:10.1088/0957-0233/18/12/017.

102. Jasiuniene, E.; Raisutis, R.; Sliteris, R.; Voleisis, A.; Jakas, M. Ultrasonic NDT of wind turbine blades using contact pulse-echo immersion testing with moving water container. Ultrasound 2008, 63, 28–32.

103. Giurgiutiu, V.; Cuc, A. Embedded non-destructive evaluation for structural health monitoring, damage detection, and failure prevention. Shock Vib. Dig. 2005, 37, 83–105.

104. Tsai, C.-S.; Hsieh, C.-T.; Huang, S.-J. Enhancement of damage-detection of wind turbine blades via CWT-based approaches. IEEE Trans. Energy Convers. 2006, 21, 776–781.

105. Drewry, M.A.; Georgiou, G.A. A review of NDT techniques for wind turbines. Insight-Non-Destr. Test. Cond. Monit. 2007, 49, 137–141.

106. Cheng, L.; Tian, G.Y. Comparison of nondestructive testing methods on detection of delaminations in composites. J. Sens. 2012, 2012. doi: org/10.1155/2012/408437.

107. Jüngert, A. Damage Detection in Wind Turbine Blades Using Two Different Acoustic Techniques. Proceedings of the 7th fib Ph.D. Symposium, Stuttgart, Germany, 11–13 September 2008.

108. Beattie, A.G.; Rumsey, M. Non-Destructive Evaluation of Wind Turbine Blades Using an Infrared Camera. SAND98-2824C; Sandia National Laboratory: Albuquerque, NM, USA, 1998.

109. Liu, W.; Tang, B.; Jiang, Y. Status and problems of wind turbine structural health monitoring techniques in China. Renew. Energy 2010, 35, 1414–1418.

110. Bodil, A.; Mats, D.; Magnus, U. Feasibility Study of Thermal Condition Monitoring and Condition Based Maintenance in Wind Turbines. Elforsk Report 11:19; ELFORSK: Stockholm, Sweden, 2011.

111. Ge, Z.; Du, X.; Yang, L.; Yang, Y.; Li, Y.; Jin, Y. Performance monitoring of direct air-cooled power generating unit with infrared thermography. Appl. Therm. Eng. 2011, 31, 418–424.

112. Yang, W.; Tavner, P.J.; Crabtree, C.J.; Wilkinson, M. Cost-effective condition monitoring for wind turbines. IEEE Trans. Ind. Electron. 2010, 57, 263–271.

113. Yang, W.; Jiang, J.; Tavner, P.; Crabtree, C. Monitoring Wind Turbine Condition by the Approach of Empirical Mode Decomposition. Proceedings of the International Conference on Electrical Machines and Systems, ICEMS 2008, Wuhan, Hubei, China, 17–20 October 2008; pp. 736–740.

114. Yang, W.; Tavner, P.; Crabtree, C.; Wilkinson, M. Research on a Simple, Cheap but Globally Effective Condition Monitoring Technique for Wind Turbines. Proceedings of the 18th International Conference on Electrical Machines, ICEM 2008, Vilamoura, Portugal, 6–9 September 2008; pp. 1–5.

115. Wilkinson, M.R.; Spinato, F.; Tavner, P.J. Condition Monitoring of Generators & Other Subassemblies in Wind Turbine Drive Trains. In Proceedings of the IEEE International Symposium on Diagnostics for Electric Machines, Power Electronics and Drives, SDEMPED 200, Cracow, Poland, 6–8 September 2007; pp. 388–392.

116. Costinas, S.; Diaconescu, I.; Fagarasanu, I. Wind Power Plant Condition Monitoring. Proceedings of the 3rd WSEAS International Conference on Energy Planning, Energy Saving, Environmental Education (EPESE '09), Canary Islands, Spain, 1–3 July 2009; pp. 71–76.

117. Fuchs, E.; Masoum, M.A.S. Power Quality in Power Systems and Electrical Machines; Burlington, MA, USA, 2008.

118. Yang, W.; Tavner, P.J.; Crabtree, C.J. An Intelligent Approach to the Condition Monitoring of Large Scale Wind Turbines. Proceedings of the European Wind Energy Conference and Exhibition (EWEC2009), Marseille, France, 16–19 March 2009.

119. Cook, K. A Power Quality Monitoring System for a 20 kW Ocean Turbine. Master's Thesis, Florida Atlantic University, Dania Beach, FL, USA, August 2010.

120. Qiao, W. Recovery Act: Online Nonintrusive Condition Monitoring and Fault Detection for Wind Turbines. Award Number DE-EE0001366; University of Nebraska–Lincoln: Lincoln, NE, USA, 2012.

121. Watson, S.J.; Xiang, B.J.; Yang, W.; Tavner, P.J.; Crabtree, C.J. Condition monitoring of the power output of wind turbine generators using wavelets. IEEE Trans. Energy Convers. 2010, 25, 715–721.

122. Amirat, Y.; Choqueuse, V.; Benbouzid, M. Condition Monitoring of Wind Turbines Based on Amplitude Demodulation. Proceedings of the 2010 IEEE Energy Conversion Congress and Exposition (ECCE), Atlanta, GA, USA, 12–16 September 2010; pp. 2417–2421.

123. Wakui, T.; Yokoyama, R. Wind speed sensorless performance monitoring based on operating behavior for stand-alone vertical axis wind turbine. Renew. Energy 2013, 53, 49–59.

124. Yang, W.X.; Tavner, P.J.; Crabtree, C.J. Bivariate empirical mode decomposition and its contribution to wind turbine condition monitoring. J. Sound Vib. 2011, 330, 3766–3782.

125. Yang, W.X.; Tavner, P.J.; Court, R. An online technique for condition monitoring the induction generators used in wind and marine turbines. Mech. Syst. Signal Process. 2012, 38, 103–112.

126. Gill, S.; Stephen, B.; Galloway, S. Wind turbine condition assessment through power curve copula modeling. IEEE Trans. Sustain. Energy 2012, 3, 94–101.

127. Yazidi, A.; Henao, H.; Capolino, G.; Artioli, M.; Filippetti, F.; Casadei, D. Flux Signature Analysis: An Alternative Method for the Fault Diagnosis of Induction Machines. Proceedings of the 2005 IEEE Russia Power Tech, St. Petersburg, Russia, 27–30 June 2005; pp. 1–6.

128. Yazidi, A.; Capolino, G.; Filippetti, F.; Casadei, D. A New Monitoring System for Wind Turbines with Doubly-Fed Induction Generators. Proceedings of the IEEE Mediterranean Electrotechnical Conference, MELECON 2006, Malaga, Spain, 16–19 May 2006; pp. 1142–1145.

129. Douglas, H.; Pillay, P.; Barendse, P. The Detection of Interturn Stator Faults in Doubly-Fed Induction Generators. Proceedings of the Conference Record of the 2005 Industry Applications Conference, Fourtieth IAS Annual Meeting, Kowloon, Hong Kong, 2–6 October 2005; Volume 2, pp. 1097–1102.

130. Yang, W.; Tavner, P.; Wilkinson, M. Wind Turbine Condition Monitoring and Fault Diagnosis Using Both Mechanical and Electrical Signatures. Proceedings of the IEEE/ASME International Conference on Advanced Intelligent Mechatronics, AIM 2008, Xian, Shannxi, China, 2–5 July 2008; pp. 1296–1301.

131. Butler, S.; O'Connor, F.; Farren, D.; Ringwood, J.V. A Feasibility Study into Prognostics for the Main Bearing of A Wind Turbine. Proceedings of the 2012 IEEE International Conference on Control Applications (CCA), Dubrovnik, Croatia, 3–5 October 2012; pp. 1092–1097.

132. Chen, B. Survey of Commercially Available SCADA Data Analysis Tools for Wind Turbine Health Monitoring; School of Engineering and Computing Sciences, Durham University: Durham, UK, 2010.

133. Schlechtingen, M.; Santos, I.F. Condition Monitoring with Ordinary Wind Turbine SCADA Data—A Neuro-Fuzzy Approach. Proceedings of the European Wind Energy Association (EWEA) Offshore 2011, Amsterdam, The Netherlands, 29 November–1 December 2011.

134. Schlechtingen, M.; Santos, I.F.; Achiche, S. Wind turbine condition monitoring based on SCADA data using normal behavior models: Part 1: System description. Appl. Soft Comput. 2013, 13, 259–270.

135. Li, H.; Hu, Y.; Yang, C.; Chen, Z.; Ji, H.; Zhao, B. An improved fuzzy synthetic condition assessment of a wind turbine generator system. Int. J. Electr. Power Energy Syst. 2013, 45, 468–476.

136. Yang, W.X.; Court, R.; Jiang, J.S. Wind turbine condition monitoring by the approach of SCADA data analysis. Renew. Energy 2013, 53, 365–376.

137. Crabtree, C.J. Survey of Commercially Available Condition Monitoring Systems for Wind Turbines; Durham University: Durham, UK, 2010.

138. Wang, W. An intelligent system for machinery condition monitoring. IEEE Trans. Fuzzy Syst. 2008, 16, 110–122.

139. Amirat, Y.; Choqueuse, V.; Benbouzid, M. Wind Turbines Condition Monitoring and Fault Diagnosis Using Generator Current Amplitude Demodulation. Proceedings of the 2010 IEEE International Energy Conference and Exhibition (EnergyCon), Manama, Bahrain, 18–22 December 2010; pp. 310–315.

140. Zimroz, R.; Urbanek, J.; Barszcz, T.; Bartelmus, W.; Millioz, F.; Martin, N. Measurement of instantaneous shaft speed by advanced vibration signal processing-application to wind turbine gearbox. Metrol. Meas. Syst. 2011, 18, 701–712.

141. Ye, X.; Yan, Y.; Osadciw, L.A. Learning Decision Rules by Particle Swarm Optimization (PSO) for Wind Turbine Fault Diagnosis. Proceedings of the Annual Conference of the Prognostics and Health Management Society 2010, Portland, OR, USA, 10–16 October 2010.

142. Boyle, D.; Magno, M.; O'Flynn, B.; Brunelli, D.; Popovici, E.; Benini, L. Towards Persistent Structural Health Monitoring through Sustainable Wireless Sensor Networks. Proceedings of the 2011 Seventh International Conference on Intelligent Sensors, Sensor Networks and Information Processing (ISSNIP), Adelaide, SA, Australia, 6–9 December 2011; pp. 323–328.

143. Christensen, J.J.; Andersson, C.; Gutt, S. Remote Condition Monitoring of Vestas Turbines. Proceedings of the of European Wind Energy Conference and Exhibition (EWEC2009), Marseille, France, 16–19 March 2009.

144. Miguelanez, E.; Lane, D. Predictive Diagnosis for Offshore Wind Turbines Using Holistic Condition Monitoring. Proceedings of the OCEANS 2010, Seattle, WA, USA, 20–23 September 2010; pp. 1–7.

145. Higgs, P.A.; Parkin, R.; Jackson, M.; Al-Habaibeh, A.; Zorriassatine, F.; Coy, J. A Survey on Condition Monitoring Systems in Industry. Proceedings of the ASME 7th Biennial Conference on Engineering Systems Design and Analysis, Manchester, UK, 19–22 July 2004; Volume 3, pp. 163–178.

146. Yang, B.; Sun, D. Testing, inspecting and monitoring technologies for wind turbine blades: A survey. Renew. Sustain. Energy Rev. 2013, 22, 515–526.

147. Lynch, J.P.; Loh, K.J. A summary review of wireless sensors and sensor networks for structural health monitoring. Shock Vib. Dig. 2006, 38, 91–130.

148. Smarsly, K.; Law, K.H.; Hartmann, D. Implementing a Multiagent-Based Self-Managing Structural Health Monitoring System on a Wind Turbine. Proceedings of the 2011 NSF Engineering Research and Innovation Conference, Atlanta, GA, USA, 4–7 January 2011.

149. Diamanti, K.; Soutis, C. Structural health monitoring techniques for aircraft composite structures. Prog. Aerosp. Sci. 2010, 46, 342–352.

150. Giurgiutiu, V.; Zagrai, A.; Bao, J.J. Piezoelectric wafer embedded active sensors for aging aircraft structural health monitoring. Struct. Health Monit. 2002, 1, 41–61.

151. Seah, W.K.; Eu, Z.A.; Tan, H.-P. Wireless Sensor Networks Powered by Ambient Energy Harvesting (WSN-HEAP)-Survey and Challenges. Proceedings of the 1st International Conference on Wireless Communication, Vehicular Technology, Infor-

mation Theory and Aerospace & Electronic Systems Technology, Wireless VITAE 2009, Aalborg, Denmark, 17–20 May 2009; pp. 1–5.

152. Eu, Z.A.; Tan, H.-P.; Seah, W.K. Design and performance analysis of MAC schemes for wireless sensor networks powered by ambient energy harvesting. Ad Hoc Netw. 2011, 9, 300–323.

153. Fu, T.S.; Ghosh, A.; Johnson, E.A.; Krishnamachari, B. Energy-efficient deployment strategies in structural health monitoring using wireless sensor networks. Struct. Control Health Monit. 2013, 20, 971–986.

154. Park, G.; Rosing, T.; Todd, M.D.; Farrar, C.R.; Hodgkiss, W. Energy harvesting for structural health monitoring sensor networks. J. Infrastruct. Syst. 2008, 14, 64–79.

155. Ling, Q.; Tian, Z.; Yin, Y.; Li, Y. Localized structural health monitoring using energy-efficient wireless sensor networks. IEEE Sens. J. 2009, 9, 1596–1604.

156. Daliri, A.; Galehdar, A.; Rowe, W.S.; Ghorbani, K.; John, S. Utilising microstrip patch antenna strain sensors for structural health monitoring. J. Intell. Mater. Syst. Struct. 2012, 23, 169–182.

157. Daliri, A.; Galehdar, A.; John, S.; Rowe, W.; Ghorbani, K. Circular Microstrip Patch Antenna Strain Sensor for Wireless Structural Health Monitoring. Proceedings of the World Congress on Engineering, WCE 2010, London, UK, 30 June–2 July 2010; Volume II, pp. 1173–1178.

158. Daliri, A.; Galehdar, A.; John, S.; Rowe, W.S.T.; Ghorbani, K. Slotted Circular Microstrip Patch Antenna Application in Strain Based Structural Health Monitoring. Proceedings of the AIAC14 Fourteenth Australian International Aerospace Congress, Melbourne, Australia, 28 February–3 March 2011.

159. Taylor, S.G.; Farinholt, K.M.; Flynn, E.B.; Figueiredo, E.; Mascarenas, D.L.; Moro, E.A.; Park, G.; Todd, M.D.; Farrar, C.R. A mobile-agent-based wireless sensing network for structural monitoring applications. Meas. Sci. Technol. 2009, 20, doi:10.1088/0957-0233/20/4/045201.

160. Park, J.-Y.; Kim, B.-J.; Lee, J.-K. Development of Condition Monitoring System with Control Functions for Wind Turbines. World Acad. Sci. Eng. Technol. 2011, 5, 269–274.

161. Isko, V.; Mykhaylyshyn, V.; Moroz, I.; Ivanchenko, O.; Rasmussen, P. Remote Wind Turbine Generator Condition Monitoring with Mita-Teknik's WP4086 System. Proceedings of the 2010 European Wind Energy Conference (EWEC) & Exhibition, Warsaw, Poland, 20–23 April 2010.

162. Koutroulis, E.; Kalaitzakis, K. Development of an integrated data-acquisition system for renewable energy sources systems monitoring. Renew. Energy 2003, 28, 139–152.

163. Feuchtwang, J.; Infield, D. The Offshore Access Problem and Turbine Availability-Probabilistic Modelling of Expected Delays to Repairs. Procedings of the European Offshore Wind (EOW) Conference & Exhibition, Stockholm, Sweden, 14–16 September 2009.

164. Karyotakis, A. On the Optimisation of Operation and Maintenance Strategies for Offshore Wind Farms. Ph.D. Thesis, University College London (UCL), London, UK, 2011.

165. Byon, E.; Ding, Y. Season-dependent condition-based maintenance for a wind turbine using a partially observed Markov decision process. IEEE Trans. Power Syst. 2010, 25, 1823–1834.

166. Jensen, F.M.; Falzon, B.; Ankersen, J.; Stang, H. Structural testing and numerical simulation of a 34 m composite wind turbine blade. Compos. Struct. 2006, 76, 52–61.

*There is one table that is not available in this version of the article. To view this additional information, please use the citation on the first page of this chapter.*

# CHAPTER 13

# Overview of Problems in Large-Scale Wind Integrations

XIAOMING YUAN

## 13.1 INTRODUCTION

Wind power installation has been growing at a rapid pace over the last decade in most countries and regions due to technology maturity and cost competitiveness of turbines among other technologies. Targets of Europe and the US in 2007 for wind installation will meet 20 % of their electricity consumptions by 2030, reaching 180 and 305 GW, respectively [1, 2]. The "Wind Power Development Roadmap" proposed jointly by the National Development and Reform Committee (NDRC) of China and the International Energy Agency (IEA) also set forth wind installation targets for China to reach 200, 400, and 1,000 GW by 2020, 2030, and 2050, respectively. China wind power generation will meet 17 % of its electricity consumption by 2050. By the end of 2010, the total worldwide wind

*Overview of Problems in Large-Scale Wind Integrations.* © *Yuan X.* Journal of Modern Power Systems and Clean Energy *1,1 (2013), DOI 10.1007/s40565-013-0010-6. Licensed under a Creative Commons Attribution License, http://creativecommons.org/licenses/by/3.0/.*

power installation reached 170 GW, with 44 GW for China exceeding 41 GW of the US as the largest installation by a country in the world.

With wind generation technology maturing and the cost decreasing close to grid parity, grid integration of wind on a large scale became a principal barrier for further deployment and has been receiving intensive attention across academia, industry, and government sectors. "Wind Monitoring 2011" [3] from China State Electricity Regulatory Commission (SERC) says that while utilization time of wind was 2,047 h in the year of 2010, it was 1,252 h for the first 6 months in 2011. Over Jan. to Aug. in 2011 in China, there were 193 wind farm tripping events, including 12 events individually losing power >500 MW. In Europe and the US, however, wind utilization was a more favorable situation due to close to load center deployment of wind and the grid interfacing performance of the turbine. In spite of this, it is expected that integration issues will aggravate with upcoming large-scale development in remote rich wind areas, such as the great plains area in the US and offshore wind areas in Europe [4].

This paper reviews differences between wind and conventional generations, provides an outline over the scope of grid integration problems and the inherent associations, and discusses thoughts toward potential solutions and relevant areas to be researched.

## 13.2 DIFFERENCES BETWEEN WIND AND CONVENTIONAL POWER GENERATIONS

1.  In terms of primary fuel and static characteristics of output power, conventional generation relying on coal or gas is constant and deterministic, while wind generation relying on wind is fluctuating and stochastic.
2.  In terms of equipment control and dynamic characteristics of output power, conventional generation is designed to mitigate and ride through grid disturbances, while wind generation is mostly designed not to mitigate grid disturbances and sometimes not to ride through grid disturbances.

## 13.3 BASIC PROBLEMS IN LARGE-SCALE WIND INTEGRATION

### 13.3.1 SUPPLY ADEQUACY

The fluctuating and stochastic nature of wind power imposes new challenges to supply adequacy in electric power systems [5, 6]. Load fluctuation or generation failure also causes a supply adequacy problem in conventional power systems. Based on static characteristics of load, the process of secure supply adequacy involves load forecasting, generation and grid planning, generation dispatch, etc. Planned generation usually consists of three types: on-line fast response generations (minutes) for frequency regulation, fast response generations such as hydro or gas for load following (hour), and slow response generations such as coal or nuclear for unit commitment (day). With large-scale wind generation in the system, the net load of the system exhibits two new features: increasing rate and range of variation; and increasing uncertainty of rate and range of variation. The study shows that these features will first impact load following requirements (increase), further on base load requirements (decrease) when the penetration grows, and yet at load level and on–off operation of conventional generations. Large-scale wind integration therefore requires a substantial increase for flexibility of response from conventional generations [1].

### 13.3.2 OPERATION STABILITY

The inability of wind turbines to mitigate or ride through grid disturbances poses new challenges to operation stability in electric power systems [7]. Small signal or large signal disturbances also cause stability problems in conventional electric power systems. Based on dynamic characteristics of the load, the process to secure operation stability involves disturbance identification, dynamic control of generations, special protection schemes for system security, etc. Dynamic control of generations responds to disturbances in three ways: active power responses to mitigate phase disturbance, active power responses to mitigate frequency disturbance, and reactive power responses to mitigate voltage disturbance. With large-scale wind generation in the system, the net dynamics of generations exhibits two new features: decreasing ability to mitigate disturbances and decreasing ability to survive disturbances. The study shows that these features first impact voltage stability at the interconnection and further impact angle and frequency stability of the system when the penetration level grows. Large-scale integration of wind therefore requires a substantial increase of the ability of wind plants to mitigate and survive grid disturbances.

## 13.4 RELATIONSHIPS BETWEEN SUPPLY ADEQUACY AND OPERATION STABILITY

Supply adequacy is a problem of static characteristics of wind power output determined from wind as primary fuel, in the time scale of minutes or above, while operation stability is a problem of dynamic characteristics of wind power output determined from turbine controls, in the time scale of seconds or below. Supply adequacy will be reached by regulating static power output with an objective of real time balancing of static power output, while operation stability will be reached by controlling dynamic power output with an objective of real time balancing of dynamic power output. The static power output characteristics of wind determine power flow and operation point of the system and therefore impact the dynamics of generations and stability of operation; on the other hand, the dynamics of wind power output determines the stability margin of the system and therefore imposes further constraints on power flow and operation point of system, as shown in Fig. 1.

As a result, turbine control is the primary factor that determines operation stability, while the fluctuation of wind power is the secondary factor for operation stability. Meanwhile, the fluctuation of wind power is the primary factor that determines supply adequacy, while turbine control is the secondary factor for supply adequacy. The system shows the operation stability challenges even with constant wind power output and shows the supply adequacy challenges even with a synchronous generator connecting wind to the grid.

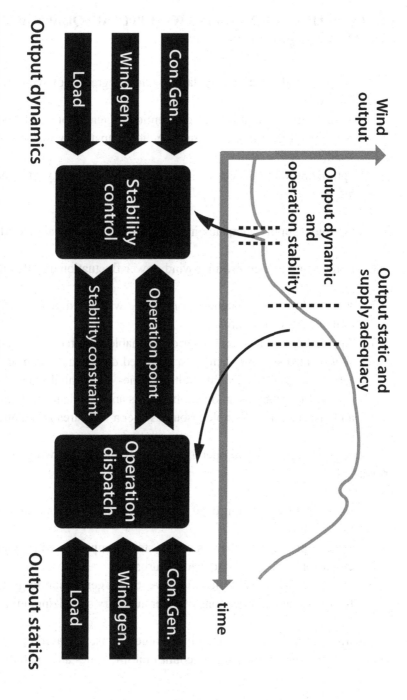

FIGURE 1: Relationship between supply adequacy and operation stability

## 13.5 CONCEPTUAL SOLUTIONS TO SUPPLY ADEQUACY AND OPERATION STABILITY PROBLEMS

1.  Fundamental solutions for large-scale integration of wind are as follows:
2.  increasing the flexibility of conventional generations in responding to wind fluctuation to improve the compatibility of the grid to wind;
3.  optimizing disturbances to improve the compatibility of wind to the grid.

Areas around supply adequacy in the following are critical as well:

1.  Forecasting and controlling wind power fluctuation and its uncertainty.
2.  Approaches for dispatching compatible with wind power fluctuation and its uncertainty.
3.  Applications of load and other dispatchable resources. Solutions to supply adequacy problem must be based on a wide area grid platform, taking into account multiple time- and spatial-scale coherences of fluctuations among wind plants and with load fluctuations, and dispatching of flexible resources over a wide area grid platform.

Meanwhile, areas around operation stability in the following are also critical:

1.  Identification and control of grid disturbances and its characteristics.
2.  Special protection schemes of a system compatible with dynamic characteristics of wind to grid disturbances.
3.  Optimization of dynamics of conventional generation to grid disturbances, and applications of other auxiliary stabilizing units.

Solutions to the operation stability problem must be based on a local wind plant platform, taking into account multiple time- and spatial-scale dynamic interactions among wind turbines, interactions with conventional generations and loads, as well as optimization of wind turbine dynamics over a local wind plant platform.

Figure 2 illustrates basic concepts for solutions to the supply adequacy and operation stability problems.

## 13.6 CRITICAL RESEARCH AREAS FOR SUPPLY ADEQUACY AND OPERATION STABILITY

Supply adequacy research shall start from the static characteristics of the output power as responding to atmospheric variation of a single turbine, moving to multiple turbines in a wind farm and further multiple farms in an area, finding interaction and smoothing mechanisms of turbine output power over a large geographical area, and identifying impact on reserve requirements of different time scales (flexibility) in the system. Forecasting and controlling (to the extent of available wind, rape rate control, for example) output power production in a wind farm or an area is important, providing basis for planning needs for reserve requirements of different time scales (flexibility) from conventional generations or other dispatchable resources. A dispatch strategy considering wind output forecasting uncertainty and wind output contingency is the grid-level tool insuring supply adequacy.

Meanwhile, operation stability research will start from the dynamic characteristics of output power as responding to grid electrical disturbances of a single turbine, moving to multiple turbines in a wind farm and further multiple farms in an area, finding interaction and aggregating mechanisms of turbine output power dynamics over especially long distance connection to grids, and identifying impact on grid stability of different voltage dimensions (phase angle, amplitude, and frequency) and time scales. Identifying features of grid disturbances and optimizing dynamics of wind turbines or farms are important to provide a basis for evaluating needs for optimizing dynamics of conventional generations or FACTS devices. Special protection systems considering stability emergencies are the grid-level countermeasures for insuring operation stability.

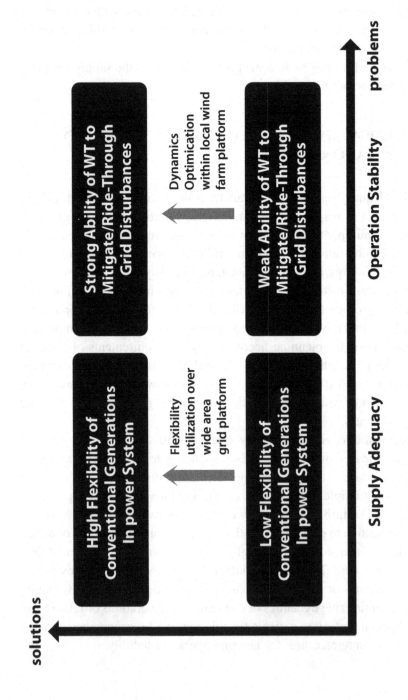

FIGURE 2: Conceptual solutions for supply adequacy and operation stability problems

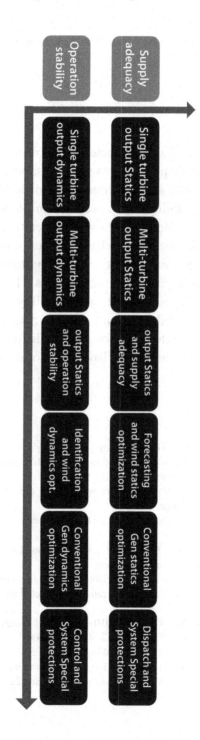

**FIGURE 3:** Critical research topics on supply adequacy and operation stability

Figure 3 shows critical research topics on supply adequacy and operation stability.

## 13.7 CONCLUSIONS

The fluctuating and stochastic characteristics of wind power output impose challenges to supply adequacy in power systems, while the weak ability of wind turbines to mitigate and ride through grid disturbances imposes challenges to operation stability in a power system. Wind power fluctuation and uncertainty determine power flow and operation point in the power system, while the grid disturbance characteristics of wind turbines impose power flow constraints and stability margins on the power system, both of which will impact limits of penetration levels. The basic solution to the supply adequacy problem is to increase flexibility in a power system in order to respond to wind fluctuations, based on a wind area grid platform, while the basic solution to operation stability is to optimize response of wind turbines to grid disturbances, based on a local wind plant platform.

## REFERENCES

1.   20 % wind energy by 2030: increasing wind energy's contribution to US electricity supply. DOE/GO-102008-2567, National Renewable Energy Laboratory, US Department of Energy, Washington, DC, USA (2008)
2.   Zervos A, Kjaer C (2008) Pure power: wind energy scenarios up to 2030. European Wind Energy Association, Brussels
3.   Kun Yang (2011) Wind monitoring 2011. State Electricity Regulatory Commission (SERC), Beijing
4.   Wiser R, Bolinger M (2011) 2010 wind technologies market report. Lawrence Berkeley National Laboratory, Berkeley
5.   Accommodating high levels of variable generation. Accommodating high levels of variable generation. North American Electric Reliability Corporation, Princeton (2008)
6.   Denholm P, Ela E, Kirby B et al (2010) The role of energy storage with renewable electricity generation. NREL/TP-6A2- 47187, National Renewable Energy Laboratory, US Department of Energy, Washington, DC, USA
7.   Xiaoming Yuan (2007) Integrating large wind farms into weak power grids with long transmission lines. Trans China Electro-Tech Soc 22(7):29–36

# Author Notes

## CHAPTER 1

### Competing Interests
The authors declare that they have no competing interests.

### Author Contributions
SOO collected the wind data and carried out initial analysis. Wind turbine calculations were performed by MSA. 'Background' section was drafted by SSP. 'Results and discussion' section was jointly drafted by MSA and SSP. All authors read, edited and eventually approved the manuscript to be submitted.

### Acknowledgments
The authors are grateful to the Nigerian Meteorological Agency (NIMET), Oshodi, Lagos, Nigeria, for providing data for this study.

## CHAPTER 2

### Acknowledgments
Financial support for this research was given by European Regional Development Fund Project 2DP/2.1.1.1.0/10/APIA/VIAA/123 (analysis of energy resource production possibilities within Latvian territorial waters and the EEZ). The authors would like to thank the Center of Processes' Analysis and Research, Ltd., University of Latvia for the CLM and Hirlam data.

## CHAPTER 3

### Conflicts of Interests
The authors do not have any conflict of interests with the content of the paper.

## CHAPTER 4

### Acknowledgments

The work presented was carried out in the scope of the interdisciplinary project "VisAsim—Visual-Acoustic Simulation for landscape impact assessment for wind farms" (2011–2014) funded by the Swiss National Science Foundation, Research Grant: CR21I2_135555. We thank Crytek for providing their game engine for free for our research purposes. In addition, we thank the community crydev.net and crysis-hq.com, namely michi.be, silent. and the_grim for providing custom 3D vegetation assets.

### Author Contributions

Adrienne Grêt-Regamey, Ulrike Wissen Hayek and Kurt Heutschi developed the overall project concept. Ulrike Wissen Hayek is managing the project and is responsible for the coordination of the development of the visual-acoustic simulation tool. Madeleine Manyoky is responsible of the development of the visualization part in the project. Madeleine Manyoky elaborated and simulated the reference situations visually and linked them to the simulated acoustics. Kurt Heutschi and Reto Pieren are responsible for the auralization part in the project. Both authors developed the emission synthesizer, the software AuraPRO for acoustically modeling all relevant propagation effects, the environmental noise synthesizer and the software RePRO for mapping the audio signals to a reproduction system. Madeleine Manyoky elaborated together with Reto Pieren the concept for linking the visual and acoustic simulation, relevant for the correct and synchronized visual and acoustic integration. All authors discussed the development and the results of the visual-acoustic simulation tool at all stages. All authors were involved in the writing process according to their project contribution and commented on the manuscript.

### Conflicts of Interest

The authors declare no conflict of interest.

## CHAPTER 5

### Acknowledgments

The authors would like to thank the following for their help with additional energy information on Ghana: Mr. E. Nketsia-Tabiri, and Mr. I. Edjekumhene.

## CHAPTER 6

### Competing Interests
The authors declare that they have no competing interests.

### Author Contributions
The manuscript was drafted by HB and revised and finalized together with KB and PT. All authors have read and approved the final manuscript.

### Acknowledgments
HB is grateful to UMCES faculty and members of the UMCES Offshore Wind focus group for many interesting discussions. KB and PT acknowledge the facilitation of their collaboration through the Marine Collaboration Research Forum (MarCRF) in Aberdeen. We thank the many colleagues in academia, industry and government who informed our understanding of these issues, and Finlay Bennet, Ian Davies and Nancy McLean for comments that greatly improved the manuscript. This is Contribution 4956 of the University of Maryland Center for Environmental Science, Chesapeake Biological Laboratory.

## CHAPTER 7

### Competing Interests
The authors declare that they have no competing interests.

### Author Contributions
MG collected the survey results, performed the interviews, and undertook the preliminary economic modeling analysis. MG also completed a draft form of sections of the manuscript. JSG conceived of the study, organized its design and coordination, and finalized the economic modeling and results analysis JSG also prepared the preliminary draft of the complete manuscript and performed revisions to produce the final draft, and also edited, read and approved the final manuscript. All authors read and approved the final manuscript.

### Acknowledgement
The authors would like to acknowledge the support of Dr. Steve Stadler and the Oklahoma Department of Commerce, and the US Department of Energy.

## CHAPTER 8

### Funding

OSU Institute for Energy and the Environment 2009 Seed grant. NASA-ROSES-2010-Climate and Biological Response grant #NNX11AP61G. RAFLES representation of buildings was developed with funding from U.S. Department of Agriculture-National Institute for Food & Agriculture (NIFA) - Air Quality grant CSREES-OHOR-2009-04566 and additional improvements to RAFLES were funded in part by NSF grants #DEB0911461 and #DEB0918869, data analysis for the campus 3-D map and eddy flux observations at the ORWRP were funded in part by NSF grant CBET-1033451, and The U.S. Geological Survey through the Ohio Water Resources Center grant #G11AP20099. Simulations were conducted at The Ohio Supercomputer Center grant #PAS0409-2. Any opinions, findings, and conclusions or recommendations expressed in this material are those of the author and do not necessarily reflect the views of the National Science Foundation. The funders had no role in study design, data collection and analysis, decision to publish, or preparation of the manuscript.

### Competing Interests

Gil Bohrer is a PLOS ONE Editorial Board member. This does not alter the authors' adherence to all the PLOS ONE policies on sharing data and materials. Gil Bohrer did not participate in any way in the editorial process of this manuscript.

### Acknowledgments

We thank Alex Hughes, Perry Livingston, Andrea Wagner and Kaleen Percha for conducting the bird surveys, Han Zhang and Yan Lu for advice on statistical analysis, Robert Beckers from Solasity.com provided advice for the analysis of wind power potential, Brady Hardiman, Kyle Maurer and Kevin Meyer for participating in collecting LIDAR data from vegetation in OSU campus.

### Author Contributions

Conceived and designed the experiments: GB PSC. Performed the experiments: KZ RLJ. Analyzed the data: KZ GB. Contributed reagents/materials/analysis tools: GB. Wrote the paper: GB KZ PSC.

## CHAPTER 9

### Funding

This research was funded by the Land Change Science Program at the United States Geological Survey. http://www.usgs.gov/climate_landuse/lcs/. Program managers had no role in the study design, data collection and analyses, decision to publish, or preparation of the manuscript.

### Competing Interests

The authors have declared that no competing interests exist.

### Acknowledgments

We appreciate the constructive comments from Diane Stephens, Steve Garman, Janet Slate, and anonymous reviewers. These greatly improved the manuscript. Abra Fowler assisted with digitizing a few of the facilities. Any use of trade, product, or firm names is for descriptive purposes only and does not imply endorsement by the U.S. Government.

### Author Contributions

Conceived and designed the experiments: JED RC. Performed the experiments: JED RC. Analyzed the data: JED. Wrote the paper: JED. Mapping: RC.

## CHAPTER 10

### Author Contributions

All authors contributed in varying degrees to writing, editing, and reviewing this manuscript.

### Conflict of Interest Statement

In terms of competing interests (financial and non-financial), the authors work for a consulting firm and have worked with wind power companies. The authors are actively working in the field of wind turbines and human health. Although we make this disclosure, we wish to reiterate that as independent scientific professionals our views and research are not influenced by these contractual obligations. The authors are environmental health scientists, trained and schooled, in the evaluation of potential risks and health effects of people and ecosystems through their exposure to environmental issues such as wind turbines.

## Acknowledgments

We thank the reviewers of this manuscript for their comments.

## CHAPTER 11

## Acknowledgments

The author would like to acknowledge the helpful comments of two anonymous referees, the domain editor and Dr. Hannah Devine-Wright upon an earlier version of this text. The author would also like to acknowledge the Wildlife and Environment Society of South Africa (www.wessa.co.za) who graciously hosted the author whilst writing up this article.

## CHAPTER 12

## Author Contributions

Pierre Tchakoua Takoutsing is the main author of this work. This paper provides a further elaboration of some of the results associated to his Ph.D. dissertation. René Wamkeue and Mohand Ouhrouche have supervised the Ph.D. work and thus have supported Pierre Tchakoua Takoutsing's research in terms of both scientific and technical expertise. Fouad Slaoui-Hasnaoui, Tommy Andy Tameghe and Gabriel Ekemb participated in designing the structure of the contributions to fit them into a review of concepts and methods for WTCM. All authors have been involved in the manuscript preparation.

## Conflicts of Interest

The authors declare no conflict of interest.

# Index

Printed in the United States
by Baker & Taylor Publisher Services